George L. Catlin

Homes on the Central Railroad of New Jersey for New York Business Men

George L. Catlin

Homes on the Central Railroad of New Jersey for New York Business Men

ISBN/EAN: 9783337417703

Printed in Europe, USA, Canada, Australia, Japan

Cover: Foto ©Suzi / pixelio.de

More available books at **www.hansebooks.com**

HOMES

ON THE

CENTRAL RAIL ROAD OF NEW JERSEY

for NEW YORK BUSINESS MEN.

NEW YORK:

1873.

Entered according to Act of Congress, by GEORGE L. CATLIN, in the year 1873, in the Office of the Librarian of Congress, at Washington.

PAMRAPO!

(See Page 11.)

LOTS, PLOTS & VILLA SITES!

Or Land by the Acre.

☞ New York Business Men can secure at this point

Beautiful Scenery,

Good Churches, Schools & Stores,

Beautiful Surroundings,

AND

Moderate Taxation,

WITHIN THIRTY MINUTES OF WALL STREET,

AND REACHED BY

FIFTY PASSENGER TRAINS DAILY!

COMMUTATION $15 PER ANNUM.

☞ This property is situated in the Third Ward of the CITY OF BAYONNE, only Five minutes from Pamrapo Station.

Terms Moderate and to suit purchasers.

The constant and rapid development of the City of Bayonne renders the purchase of property at this point

A SAFE AND PROFITABLE INVESTMENT!

For Maps and full particulars address

WILLIAM CURRIE, Executor,

GREENVILLE, N. J.

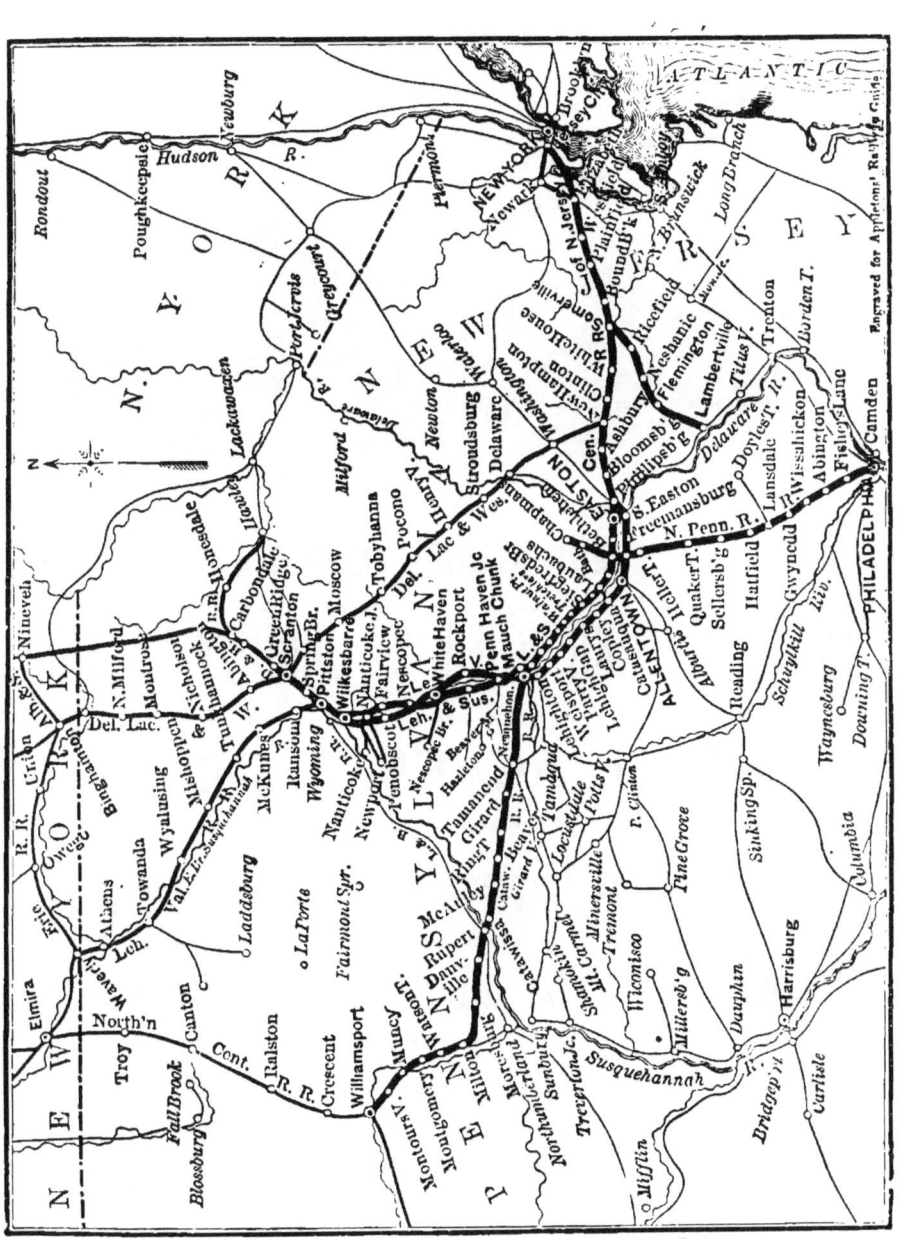

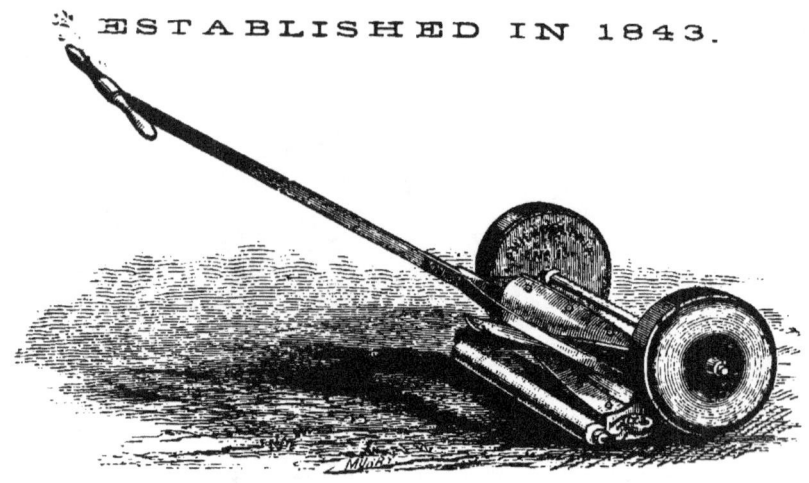

ESTABLISHED IN 1843.

THE PHILADELPHIA LAWN MOWER.

EVERY REQUISITE FOR A

Lawn, Garden, Farm or Country Estate.

R. H. ALLEN & CO.

MANUFACTURERS OF EVERY VARIETY OF

Agricultural Implements and Machinery,

AND DEALERS IN

FERTILIZERS, SEEDS AND HORTICULTURAL TOOLS.

The Allen Works, Cor. Plymouth, Jay & John Sts., Brooklyn.

Agricultural Warehouses, 189 & 191 Water St.,

P. O. Box 376.   **NEW YORK.**

HOMES

ON THE

CENTRAL RAILROAD OF NEW JERSEY

— FOR —

New York Business Men.

A DESCRIPTION OF THE REGION TRAVERSED BY THE CENTRAL RAILROAD OF NEW JERSEY, AND ITS BRANCHES AND CONNECTIONS FROM NEW YORK TO MAUCH CHUNK, EMBRACING A STATEMENT OF THE INDUCEMENTS AND CONVENIENCES HELD OUT CONJOINTLY BY THE RAILROAD COMPANY AND PROPERTY OWNERS AND OTHERS ALONG THE LINE TO THOSE DESIROUS OF SECURING

EITHER

PERMANENT OR TRANSIENT HOMES OUTSIDE OF NEW YORK.

> " Scorn not the muse because 'mid scenes like these
> She loves to wander; and with calm delight
> Prefers to dwell among the rustic homes
> Where sweet Content beside the well swept hearth
> Sits like an Angel, and will not depart."
> —T. BUCHANAN READ.

— BY —

GEORGE L. CATLIN.

PUBLISHED FOR GRATUITOUS DISTRIBUTION

BY THE

CENTRAL RAILROAD COMPANY OF NEW JERSEY.

NEW YORK:

1873.

INTRODUCTION.

While New Yorkers are jostling and crowding one another with their marble palaces and brown stone fronts, until Manhattan Island bids fair in a few years to become covered with brick and stone from end to end; while the wearisome problems of rapid transit and the East River bridge, are still unsolved, leaving Westchester County and Long Island no nearer New York, in point of time and convenience, than they were two decades ago; while thousands of people who seem never to look farther than their noses, and, worse still, who don't care to, are delving away all day long in stores and offices down town, and huddling themselves away at night in close, expensive quarters up-town; while hungry city landlords are raising rents and the deuce with their tenants at the same time. While all this is taking place, reader, remember that just across the Hudson, in fair, fertile, well governed Jersey, are thousands of dells and knolls and pretty villages, where a business man can, almost for the asking, secure a neat, convenient and healthful home, no further from his office than an up-town residence, and surrounded by scenes and influences that will give a lightness to his cares, and a halo to his domestic happiness. To a contemplation of facts and places such as these, it is that the reader's attention is invited.

G. L. C.

HOMES ON THE CENTRAL
Railroad of New Jersey.

We shall take it for granted that the reader has made up his mind to secure a home in New Jersey. The inconveniences, expenses and trials attendent upon the life in the city of any young married man dependent simply upon a salary or a moderate income, have been too often dwelt upon, and by too many experienced to require repetition here. A farewell then, once for all and forever, to exorbitant rents, foul streets, long rides in the horse cars, expensive marketing, and doctors' bills. Let us go out into the country, buy cr rent a cottage or villa in some breezy, healthful spot, remote from the city's din and dust, where we may rear about us fresh comforts and beauties as life goes on, and in old age, if Providence permits, be gladdened by the sight of our children's children playing under the leafy shadows of great trees, which, as tender saplings, we ourselves planted.

But whither shall we go? There are a score of routes radiating from the Metropolis, over each of which thousands daily pass from and to their homes. Which shall we choose as offering the most extended facilities for communication at moderate rates, and as traversing a region of continuous cities and charming villages for miles away; as landing its passengers in New York at a point within five minutes walk of the great mercantile and financial center of the city; as conducted and managed with the systematic precision of clockwork; in short, as a model railroad for local or excursion travel? Why, by all means, the Central Railroad of New Jersey, with its forty trains each way daily, with no Sunday trains to bring the rabble of Sabbath breakers, as on many other lines, to mar the quiet of your suburban home; with its palatial and fleet ferry boats, its prompt time in starting and arriving, and its numbers of beautiful landscapes, awaiting the coming of man's hand to dot them with smiling peaceful homes.

So, reader, let us together stroll down Liberty Street this bright summer morning, and we shall find the ferry-boat in waiting. A staunch, noble craft she is, handsomely as well as comfortably fitted up within. The saloons are spacious and well ventilated, lit with gas, warmed by steam, and kept as clean and neat as a good housewife's pantry. There are three others exactly like her, too. And notice how commodious is the ferry house on the wharf. Everything tells of system and a zealous regard for the comfort and safety of the traveling public. But hark! there goes the gong. Now we are off. Suppose we go out forward, catch the morning breeze and get a view of the river and bay. Here on our left are the Battery and Castle Garden, swarming with newly arrived representatives of half the nationalities of Europe; beyond there are Governor's Island and Fort William Henry; further on, Brooklyn and the wooded shores of Bay Ridge, and, in the hazy distance, Fort Lafayette and the Narrows. Then following the line of vision westward, we see the rounded blue hills of Staten Island, Robbins Reef Light,

Bedloes and Ellis Islands, and about them the rippling waters of the bay dotted with every variety of craft. A glorious spectacle this on a summer morning, before the burden and heat of the torrid day have begun. Or, if we turn about, and glance up-stream, the scene is none the less inspiring. There are the Cunard Docks, and above them those of the White Star Line. What a fleet of ferry boats, alive with people, are hurrying cityward. In the distance we can see the Bremen and Hamburg Steamers, beyond them again the Stevens Castle and Weehawken, and in the mist above, may be discerned the bold outlines of the Palisades. But we have scarcely begun to enjoy the exhilarating scenery, when again the gong sounds, our speed slackens and we are in the ferry slip at Jersey City. As we land, the depot is just before us, and see, there is the train in waiting, the locomotive at its head, with steam up and impatient for a start. "This way for Elizabeth," cries the brakeman. "All aboard" shouts the conductor, and in another moment we are off.

Nice comfortable cars these, and then they start so promptly; there is not a moment's delay. Now, we are whizzing away over a wilderness of tracks, and presently come in sight of the bay again, along the western shore of which lie the next few miles of our journey.

Now, let us consult the time table for a moment or two, for from it we may gain some interesting and valuable facts. We left the corner of Wall Street and Broadway at ten minutes before eight o'clock. Here we are at ten minutes after eight whizzing away from the Jersey City Depot. Now let us suppose that a third party, A, set out from Trinity Church at the same time with us, to take the Sixth Avenue cars to Central Park. A comparative statement of his position and our own at short intervals on our respective routes, will enable us to arrive at a clear understanding of the convenience and proximity to New York of these different stations along this portion of our line. When A, for instance, has reached the corner of Chambers Street and West Broadway, we are at Jersey City; when he is at Canal Street we are at Communipaw; as he crosses Broome Street, we reach Claremont; when he is at Clarkson Street, we are at Greenville; we pass Pamrapo, Bayonne, and Centerville, severally, as he passes Bleecker Street, Fourth Street and Waverley Place. Our arrival at Bergen Point is simultaneous with his at Ninth Street; we come to Elizabethport, as he comes to Fourteenth Street, and as our engineer whistles down brakes for Elizabeth, the slow coach horse car will have done well if it is within the shadow of Booth's Theater at Twenty-Third Street.

So here we have traveled twelve miles, while A has traversed scarcely two-thirds of the distance to Central Park. We have, moreover, had comfortable seats, a pleasant breeze, charming scenery, and a chance to smoke our cigars if we wished, while he has been sitting, or, perhaps standing, in a crowded horse-car, moving at a snail's pace over the pavements. He, too, has been jostled by the squalid, the lame, the halt and the blind. We, on the other hand, have had our choice of company. Nor have we, either, it may be remarked, during our entire ride to Elizabeth, been outside of incorporated city limits, for our route has lain through the three successive cities of Jersey City, Bayonne and Elizabeth, each of them regularly laid out in streets and avenues, and possessing all the conveniences of modern civilization.

Elizabeth, then, is no further in point of time than Twenty

third Street from the financial center of New York; and Plainfield is about equi-distant therefrom with Central Park. Let these facts, and the superior convenience and comfort of the out of town route, be borne in mind by those who are still undecided in their choice of homes. Nor should mention be omitted of a liberal system of package delivery established on this road alone of all centering in the Metropolis, by which *pater familias* may make his morning purchases in Washington Market, or the ladies do their shopping up-town, and at the ridiculously small charge of fifteen cents, find the package awaiting them at the depot at home on arrival. Can any up-town conveniences excel this?

And, now, here we are at

COMMUNIPAW AVENUE,
(15 minutes; 26 trains each way daily.)

the point at which the Newark and New York road diverges from our main line. Ascending the stairway which connects the platform with the avenue, (for our roadway passes under the grade of the latter), we shall find that we are still in the midst of the populous city, surrounded by churches, rows of brick dwellings and well graded streets. From Pacific Ave-

---

FOR HOMES IN COMMUNIPAW
APPLY TO
WOODWARD & SHERWOOD,
REAL ESTATE & INSURANCE AGENTS,
No. 15 Montgomery St., Jersey City.

☞ Particular attention given to Negotiating Loans on Bond and Mortgage in Hudson County.

R. W. WOODWARD.   T. P. SHERWOOD.

nue, a few squares west, the horse-cars run direct to the Jersey City ferry.

Communipaw, which is now the Sixth District of the city, first began to be made available for residences about sixteen years ago. It was then an open country, but has developed rapidly, and now embraces some of the most attractive building sites within the city limits. A city house, with all the modern improvements, can be rented at $500, while lots can be purchased on easy terms of payment, and within five minutes' from the depot, at prices varying from $500 to $3,000. The card of Messrs. Woodward & Sherwood will direct the reader's attention to their facilities for giving accurate information regarding property at this point.

CLAREMONT AVENUE

(17 minutes; 19 trains each way daily.)

is our next stopping place, and is the outlet for a populous section of the city, lying between our line and that of the Newark Railroad. Upon the high, wooded ridge, a short distance west of the road, may be found some admirable sites for homes, cammanding a fine view of the neighboring cities and the Bay.

A short distance beyond Claremont we cross the Morris Canal, spanned here, as at another point further on, by a substantial bridge of iron, and gain a charming view of the Bay on our left, while the green upland on our right is dotted by the marble stones and monuments of Bay Cemetery. A moment more, and we have reached the depot at

GREENVILLE,

(20 minutes ; 26 trains daily.)

where, as at Communipaw, the road passes below the level of the streets, and where an ascent of the stairway to the depot affords us at its summit a commanding view of the surrounding scenery. But, if one would gain a true idea of the natural beauty of this locality, let him stroll up Danforth Avenue, from the depot to the summit of the ridge or neck, here dividing New York and Newark Bays. For there, looking westward, he will see the blue waters and wide stretch of meadows, with distant trains creeping snail-like across their surface, the spires and chimneys of Newark, and in the blue distance, the Orange Mountains, while, turning eastward, he may see, through the leafy vista, the villa-lined shores of Long and Staten Islands, and the Narrows, alive with craft.

Hudson County Land and Improvement Co.

POST OFFICE BUILDING,

GREENVILLE.

FINE LOTS AND VILLA SITES,

AVAILABLE FOR

Suburban Residences.

Amid such scenes as this, homes at Greenville may be secured. Recently incorporated a part of Jersey City, it yet retains many of its rural beauties, while offering many of the city's conveniences. Horse-cars run direct to Montgomery Street, while stores and market wagons supply residents with the wants of daily life. The grand Boulevard projected by the Jersey City authorities, to extend from the Palisades to Bergen Point, will pass through Greenville and the adjacent city of Bayonne, materially beautifying both, and adding greatly to their direct communication with all points above and below.

There are churches of nearly all denominations, a private academy, and a first class public school, with a principal and seven assistants. Market gardening is carried on here to a considerable extent. In short, Greenville is a charming *rus in urbe*, where the New York business man may find quiet and repose at the close of his daily toil.

Property here is all of it high and desirable, and a man of salary or moderate means can buy lots at $400 each, have a house built for him, pay a small percentage down, and leave the remainder on mortgage for five years.

JAMES R. WILLIAMS,
REAL ESTATE AGENT,

POST OFFICE BUILDING,

GREENVILLE, HUDSON CO., N. J.

A ride of three minutes' more, during which we again cross the Morris Canal, forming at this point the southern boundary line of Jersey City, brings us within the limits of the new and growing city of Bayonne, in which our first stopping place is the station called

PAMRAPO,

(23 minutes ; 26 trains each way daily,)

deriving its singular title from a corruption of the old township name of Pembrepock. This is now the Third Ward of the city of Bayonne, and is regularly laid out in streets and avenues, the former, which extend across the neck, being numbered; the latter, which run lengthwise, or parallel with the bay, being lettered. Pamrapo, or the Third Ward, extends from Division Avenue, or Thirty-fifth Street, to Fifty-sixth Street, or the Jersey City line.

FOR HOMES IN PAMRAPO
APPLY TO
BRAMHALL & SEYMOUR,
Cor. Bayonne Ave. & Ave. D., or No. 1 Exchange Pl., Jersey City.

Adjacent to this depot one finds an abundance of locations eligible for homes. The ground rises westward from the railroad, and at almost any point commands a fine view of the water. There are here an Episcopal Church, a fine public school, and stores of all kinds. The Bayonne Yacht Club has its club house here. The adjacent drives, too, are pretty and varied. Half a mile from the depot is the quaint little hamlet of Saltersville, a vestige of the days when city limits hereabout were never dreamed of, while further back, on the shores of Newark Bay, are some sylvan retreats which cannot but charm the lover of natural beauties.

Upon the second page of the cover the reader will note the advertisement of the estate of Jas. Currie, deceased, calling attention to some desirable building sites, which are offered for sale by lots, plots or the acre, located on curbed and flagged streets and avenues, within five minutes walk of the depot. The property commands a glorious view of the two Bays and Narrows, and of Long and Staten Islands, is covered with a fine growth of shade trees, and has the additional advantage of horse car communication direct with the Jersey City Ferry. The purchase of any portion of this property may be safely recommended as an investment.

Not over a dozen squares beyond Centre Street, or Pamrapo station, we stop at Bayonne Avenue, where is located the central station in

BAYONNE.
(25 minutes ; 26 trains each way daily.)

Close at hand, on Avenue D, are the City Offices and the City Hall, where weekly assemble the City Fathers to discuss measures for the improvement of their growing city. Bayonne was incorporated in March, 1870, has now a population of about six thousand, with four public schools, three post offices, churches of all the leading denominations, a weekly paper, a Masonic Lodge (Bayonne No. 99), a gas company, uniformed police force, and a good system of sewerage. Its avenues stretch in a magnificent sweep from the Jersey City line to the Kill Von Kull, its flagged sidewalks extend in all directions as far as the eye can reach ; its public buildings are creditable in their size and architecture, and its private dwellings are most of them models of beauty and tastefulness. This description may be considered as applying to the entire city of Bayonne, for, throughout its whole extent, there is apparent an infusion of energy, and a spirit of improvement from which some larger and older cities might well take example.

Lots in the vicinity of the Bayonne Avenue station can be

had at prices varying from $500 to $1200. The intending purchaser can obtain accurate information as to particular localities by calling at the Real Estate Headquarters of Messrs. Bramhall & Seymour, one square from the depot.

FOR HOMES IN BAYONNE
APPLY TO
BRAMHALL & SEYMOUR,
Cor. Bayonne Ave. & Ave. D., or No. 1 Exchange Pl., Jersey City.

At
CENTERVILLE,
(28 minutes ; 26 trains each way daily.)

which is another station established at 27th Street for the convenience of residents of this portion of the city, including Constables Hook, the prices of land are about the same as those last quoted. Residents in this vicinity are within convenient distance of the churches, school house and stores.

And now, on our left, as we proceed, the view grows each moment more varied. Close at hand is the Kill Von Kull, and beyond it are the wooded hills of Staten Island, adorned with cottages and country seats. Nearer still, we presently see the company's great coaling depot and wharves at Port Johnston, whence annually are shipped hundreds of thousands of tons, brought direct from the great fields of the Lehigh Valley. As an outlet from the coal regions to the seaboard, the Central Railroad of New Jersey is the most available of all routes.

BERGEN POINT,
(31 minutes ; 32 trains each way daily.)

But here we are at what is probably the best known, as it is the most densely populated section of Bayonne, comprising the first and fourth wards of the city, and long famous as a summer resort for New Yorkers.

FOR HOMES IN BERGEN POINT
APPLY TO
BRAMHALL & SEYMOUR,
Cor. Bayonne Ave. & Ave. D., or No. 1 Exchange Pl., Jersey City.

Upon alighting at this portion of the city, the visitor is at once struck with the evidences of neatness, enterprise and liberality visible on all sides. The streets are wide, well paved, and kept in splendid condition, the sidewalks flagged and lined at frequent intervals with handsome gas lamps, while the beauty of the dwellings, public edifices, lawns and shrubbery gives assurance that here both wealth and culture abide.

Opposite the depot stands the commodious and elegant building of the Young Men's Christian Association, and if, after admiring this, the visitor will pass up Sixteenth Street toward the high ground overlooking Newark Bay, he will find it lined with costly villas, the abodes of prominent New York business men, with here and there a broad roadway, sewered and flagged, running to the water's edge, and offering attractive sites for the erection of homes. The view at this point, too, is superb. And now, let the visitor retrace his steps, and pass through the more densely settled portion of the ward. He will find Episcopal, Dutch Reformed, Roman Catholic and Lutheran churches, a brick school-house said to have cost $20,000, an Institute for young ladies, a gymnasium, and stores of all varieties. And, passing all these, he will reach the shores of the Kill Von Kull, skirted by a splendid drive, and lined with attractive residences.

· Here is that famous summer resort, the Latourette House; here, too, the Club House of the Argonauta Rowing Association, and here, on pleasant summer evenings, when the cool breeze blows in from the bay, one may find in a stroll or ride the perfection of quiet comfort and beauty.

Within the past few years Bergen Point, which, by the way, derives its name from the fact of having been originally (in 1616) settled as a trading port by some colonists of Norwegian extraction, has been annually growing in importance and popularity as a home for New Yorkers. Property is in constant demand, lots within ten minutes of the depot selling at from $1000 to $1500, and at good points overlooking Newark Bay at $1200. And yet, a little over two hundred years ago the entire section which we have traversed from the Jersey City ferry to this point, and the land covered by the present Jersey City besides, was sold by the Indians for 80 fathoms of wampum, 20 fathoms of cloth, 12 brass kettles, 6 guns, 2 blankets, 1 double brass kettle and half a barrel of strong beer. Poor Lo was a better fighter than speculator in real estate.

But the train is here and we must be off again. Now we rush through a heavy cutting, crossed at intervals by street bridges, and in another moment dash out upon the long bridge spanning Newark Bay, and the view breaks upon us in all its beauty. On the left we see the wide expanse of water stretching away into Staten Island Sound, and amid the numerous sails can discry the neat little lighthouse on Shooters Island. Here, too, the hills of Staten Island slope away into fertile fields and meadow lands, dotted with farm houses, and lined at the water's edge with successive villages. Or, looking up the bay, we see Newark, stretching out her boundaries on all sides, while beyond loom up the uplands of Morris and Passaic. Now we are fairly out on the bridge; it is about two miles long, and we may almost imagine ourselves skimming over the bay in a sail-boat, so novel is the isolation, so distant seem the shores behind and before us. Now we slowly pass the iron draw and are off again. Now the western shore grows rapidly nearer, and, almost before we know it, we are on terra firma again in the third successive city on our route—the great city of Elizabeth ; and now we dash under the shadow of the great cluster of elegant and substantial brick buildings which the Singer Manufacturing Co., have recently erected at this point, at the cost of over two millions of dollars, with a view to combining in one grand establishment all the various departments of their enormous manufacturing business, hitherto carried on at various widely separated points in this country and abroad. When completed, this mammoth structure will cover nine acres of flooring. The main building, fronting on First and Trumbull Streets, is an imposing fire-proof edifice, 1100 feet long, 50 feet deep, four stories in height, and covered with a slate mansard roof, from which rise stately towers. In magnitude the building reminds the observer of the much admired Grand Central Depot at Forty-second Street, though the latter in dimensions is much the smaller of the two. Then, in addition to the main building, we see the foundry, which, fronting on the railroad, is 600 feet long, and 100 wide. As we clatter by, we catch glimpses through the grated windows of hundreds of Vulcan's votaries, hurrying hither and thither amid the glowing furnaces. Then we see, too, the building used for cleaning castings, forging, japanning, etc., which is 530 feet long and 50 deep, with two wings of 75 by 130 feet each, and last of all, a cabinet-case shop and a box factory, each 200 feet long, 50 feet deep and three stories high. The boilers, engines, elevators, hoistways and stairways, are all outside the main buildings. About twenty miles of steam pipe are used in heating the premises, together with twenty boilers worked by engines of seventy-five horse power each. The total frontage is about 3000 feet. On the grounds which cover 32 acres, are five stationary engines, aggregating nearly 1000 horse power, and nearly two miles of railroad track, connected by switches with those of the Central R. R., of New Jersey. This, with a water frontage of more than a thousand feet, gives the Company unusual facilities for receipt of iron and coal from the West, or the shipment of material to all parts of the world.

In the construction and organization of the various works of the Company, it should be stated that great credit is due to the well directed energy of Mr. George R. McKenzie, who is identified with the manufacturing department of the business, and who superintends the construction and organization of the various factories. He may be said to be in that department what Mr. Inslee A. Hopper, the President of the Company, is in its extensive and successful commercial relations.

The Singer Company will employ upwards of three thousand men here when the works are fully in operation, and will be ready to turn out five thousand machines per week. But even this immense supply will barely serve to keep pace with the demand, as will be seen by a statement of their last year's sales as compared with those of other companies, published on the last page of this work.

Nor is the indirect advantage of this magnificent piece of

enterprise to be overlooked, for it will probably increase the population of Elizabeth by ten thousand, and its business by hundreds of thousands of dollars, annually disbursed by the Company to their employes. Here, too, the operatives will find convenient homes for their families, and amid the healthful surroundings of this attractive spot will recall brighter days, and feel that they are beginning life anew, while about the works, where now are open lots, will spring up a new settlement of neat and tasteful dwellings. Lots which two years ago sold here at $250 now command, in some instances, from $1000 to $2000.

So then as we whiz by this little city on the shore, we cannot but moralize a little on the fact that the Singer Manufacturing Company are true philanthropists in this, that while enhancing their own interests, they do fully as much, if not more, good to all about them. But we are startled from our reveries by the harsh rattle of the brakes, and in a moment more, have reached the station at

ELIZABETHPORT,
(38 min. 32 trains each way daily.)

the first of the four depots established by the company within the limits of this great and growing city. Now, here we shall alight, for aside from the claims presented by this portion of the city as a place for residence, there is much more of historic and local interest to entertain and instruct the visitor. North of the track, the meadows stretch away in an almost unbroken sweep to the city of Newark in the distance. But, turn about, and the scene is one of life and improvement. Here is the horse car in waiting to take us if we please, to the other end of the city. But perhaps we had better walk. Before us are innumerable cottages and dwellings, the homes of the humble; but beyond these simple abodes, we shall find business streets with handsome stores and offices, a public park, and then dwellings on all sides, combining every element of luxury and taste. On our left, we see the immense coal wharves which draw to Elizabethport five million tonnage annually, and give it a prominence as a commercial point, which promises ere long to establish it as an independent port of entry. A visit to one of these, and an inspection of the rapid and systematic manner in which vessels are loaded will repay the observer.

Elizabethport comprises the first three wards of the city, yet retains its own name and Post Office. Its growth within the past few years has been marvelous. The visitor of a decade ago, will remember it as the point at which, after a tedious steamboat sail from New York, he disembarked to take the cars for Elizabeth and points beyond it. Now we are whisked hither in less time than it would take Puck to put a girdle round the earth, and lo, we find the quaint old Elizabethport of the past replaced by a great active, bustling city, full of life and industry, every day becoming more and more a manufacturing and commercial center, and possessing a water front or dockage of one and a half miles, with all the prolific coal, iron and lumber regions of Pennsylvania at its back, and directly connected with it by rail.

With such natural advantages, with good churches, schools and stores, and with an enterprising people to avail themselves of them, it is not difficult to prophesy for this section of the city of Elizabeth, a substantial and permanent commercial prosperity. And with it, too, will come hundreds of new seekers after homes, tired of the Metropolis, and anxious to secure a quiet retreat in one of the many pleasant streets in which this portion of the city abounds. Already

---

HOMES ON THE CENTRAL AT ELIZABETHPORT.

ROPES & POTTER,
Real Estate and Insurance Offices,
At CENTRAL RAILROAD DEPOT,
AND
No. 50 FIRST STREET, ELIZABETHPORT, N. J.

Houses and Lots in Every part of ELIZABETHPORT and ELIZABETH. Villa and Villa Sites, Building Lots, Farms, Manufacturing Property, and Land by the acre throughout Central and Eastern New Jersey.

☞ **Ask for Ropes' Real Estate Register.**

ELIHU H. ROPES.   GEO. N. POTTER,

---

there are eight hundred daily travelers from this point to New York, and three thousand from the entire city. What will be the demand for transportation at no distant day when all these available sites for homes are occupied by men doing business in the Metropolis? And in this connection the reader's attention is directed to the foregoing card of Messrs. Ropes & Potter, a firm dealing largely in real estate at this point, and able therefore to give accurate and reliable information to purchasers regarding it.

SPRING STREET STATION
(40 min.   21 trains each way daily.)

is the depot established for the convenience of those residing in the populous section lying midway between "the port" and the original business center of the city. Within a square or two about it may be found for rent good dwellings with all the city conveniences, at four or five hundred dollars per annum, while lots can be purchased at from $350 to $750.

And now, as we proceed, the great brick blocks close in

more closely about us, the indications of our approach to the city's center rapidly multiply, and almost before we know it, we cross the track of the New Jersey R. R., which here intersects with our own, and are at the depot at

ELIZABETH.
(36 min. by Express trains, 45 min, by Acc. 42 trains each way daily.)

And now, as we alight, and slowly stroll up the shaded sidewalks of Broad Street, past the long rows of store windows, in their display rivaling those of the Metropolis, past the brick and brown stone rows of dwellings, past Library Hall, the churches, the Court House, and the pleasant lawns to the quieter beauties of the hill beyond, let us recall a few of the incidents and traditions which have combined to make this ground on which we tread, historic.*

The first white settlers, it seems, were three adventurous Long Islanders, Bayley, Denton and Watson, who, in October, 1664, effected the purchase of some four hundred thousand acres lying between the Raritan and Passaic Rivers, for a petty consideration in the shape of a few guns, kettles, and other articles precious to the aboriginal heart. They established a settlement forthwith, but, before they had enjoyed possession for a twelve month, Lord Berkeley and Sir George Carteret, under the grant of the Duke of York, assumed control as Lord Proprietors. Two years later, the village which had increased to four score families in population, was called Elizabeth Town, in honor of Sir George's beautiful wife. It enjoys the name of having been the third settlement made in New Jersey, and the *first* by the English.

---

* The author is indebted to Mr. E. H. Ropes' admirable little pamphlet entitled "Elizabeth in Olden Time," for much of the information here given.

The surrender of New York to the Dutch in 1673, also brought Elizabethtown temporarily under the sway of the Netherlands, but, in the following year, English rule was restored, and Carteret reinstated. Thenceforward, the place grew and prospered. In 1680, there were seven hundred inhabitants, and thirty thousand acres under cultivation, and glowing accounts of the fertility of the soil and healthfulness of surroundings of this modern Canaan, were sent back to the Old World by the settlers. In 1703, New Jersey became a Royal Province, and in 1740 the "Free Borough and Town of Elizabeth" was incorporated under a charter from King George the Second. Twenty-four years later, Elizabeth celebrated its centennial by a grand public barbecue in the center of the town.

But it was when the troublous times that tried men's souls dawned upon the American Colonies, that the most heroic pages of the history of Elizabeth were recorded. As early as February, 1766, the people, it is stated, threatened to hang without Judge or Jury, any one giving adherence to the odious Stamp Act. And when the call to arms came ringing from Lexington and Concord, this plucky little borough sent not only a large supply of powder to the front, but followed it with *sixteen companies of infantry and one of cavalry*. The latter company, by the way, served as Lady Washington's escort on a portion of her journey to join her husband at Cambridge.

As might have been anticipated, such a display of patriotism provoked no small hostilities on the part of the enemy, who held possession of New York and Staten Island, lying directly opposite. On the night of the day on which the Declaration of Independence was signed, a British sloop of fourteen guns appeared off the town, but was attacked by the citizens, armed with two howitzers, and, after losing several of her crew, was fired and destroyed. This exploit, occurring within three hours after the birth of the United States of America, is justly claimed to have been the first in its military annals.

Here, in these memorable days, dwelt Governor Livingston, the first Chief Magistrate of the State; here, that fiery patriot Rev. James Caldwell, pastor of the old Presbyterian church, from whose congregation went forth five Generals, three Colonels, five Majors and a host of subaltern officers, to take commands in the Continental army. Mr. Caldwell by his fervor and zeal, became so odious to the enemy that during his preaching it was necessary to post sentinels about the church, and to keep his own pistols on the pulpit beside him, to prevent, if possible, his surprise and capture. In fact, in 1779, the British did essay to surprise the place by crossing a detachment from Staten Island, but were repulsed with severe loss. The good old church was however fired in the following year by the torch of a refugee. Mrs. Caldwell was murdered by Knyphausen's troops, and the reverend gentlemen himself was, in 1781, shot and killed by an Irish soldier of the American army, who was supposed to have been instigated to the deed by the British authorities at New York, and who was subsequently hanged for the crime.

After the close of the war, Washington en route to his inauguration at New York, passed thorough Elizabethtown, and was met at the port by a flotilla with music and artillery. The old hotel known as the Pountney House, at which he breakfasted, still stands, being on Trumbull Street, within the enclosure of the Singer factories previously spoken of.

Since the revolution the growth of Elizabeth has been steady, and of the last few years all but incredible. In 1830 her population was 3445; in 1840, 4184; in 1850, 5583; in 1855, the city was incorporated; in 1860. she had 11,567 inhabitants; in 1865, 17,373; in 1860, 20,848; and to-day has about 25,000.

But, after all, it is in the Elizabeth of to-day that the New York business man in search of a home, is most interested. These scenes and memories, 'tis true, give a flavor to the enjoyment of a residence in their midst, yet serve but poorly to supply the sound, practical information which the modern business man desires in his selection of a spot where he and his family may abide. So let us look at Elizabeth, not in history, but as we see it with our own eyes. The name has become, first and foremost among the New Jersey cities, the synonym for all that is enterprising and progressive. Take these figures for instance. The city covers an area of nearly 12 square miles, has over 69 miles of streets, about one-third of which are paved, $31\frac{1}{2}$ miles of sewers, and $85\frac{1}{2}$ miles of flagged sidewalks. There are between 90 and 100 manufacturing establishments of various kinds, and 14 coal shipping docks. Then, turning to another side of the picture, we find that there are thirty churches, (including the famous Westminster church, costing $200,000), the best of schools, both public and private, twelve hotels, three daily, one semi-weekly, three weekly and one monthly newspaper, five insurance companies, six banks, an eighty thousand dollar market, an Arcade building which cost $150,000, an Orphan Asylum costing $50,000, and hundreds of private dwellings, which in elegance of style and construction would grace any city. The ratio of taxation to the actual value of property is computed to be only four-fifths of one per cent., an exceedingly low rate for a city developing so rapidly.

Where, then, can one seeking a city home out of the city, better seek it than amid such surroundings as these. Such a one will find wide shaded avenues, paved and underlaid with gas and water pipes and sewers, stretching away in all directions antennæ-like from the city's center; and on these, tasteful cottages and villas, built with every convenience of modern times, awaiting his occupancy; if he would rent, at rates far lower than those in New York; if he would purchase, on terms so moderate that no man, careful of his own and his family's future, would be justified in treating them with unconcern. And such in brief are the claims of Elizabeth as a place for a Home on the Central Railroad of New Jersey.

Upon resuming our ride westward, we pass on through the thickly settled western section of the town, cross the Elizabeth River, (which, passing through the city's center, finds an outlet at Staten Island Sound), shoot under the Cherry and Chilton Street bridges, which in turn span the track, and presently come to a standstill at the depot at

WEST ELIZABETH.
(42 min. 11 trains each way daily.)

Here we are still within the city limits, and, within a stones throw of us, on Grand Street and Westfield Avenue, may be seen rows of attractive dwellings, while here and there are interspersed grateful reminders of the old regime when Mansard roofs were unknown, pleasant country seats, embowered in foliage, fronting on shaded, close cropped lawns, and surrounded with broad verandahs, where one may sit comfortably ensconced, and watch the tide of travel pass and repass before him. Possibly the following anecdote from a recent number of *Le Journal Amusant* had its origin here.

"A friend paid a visit to a country house.

"The view from here leaves little to be desired," said he to the proprietor.

"Do you think so? It looks toward the railroad station."

"Yes, I observed that. That certainly does not add to the charm of the thing."

"Pardon me. It is very funny. We see all the people who miss the train."

We pause but a moment at West Elizabeth, and then with a shriek and a roar are off again. Now, for the first time since leaving New York, we are out in the open country, among the clover and daisies and buttercups; we whiz past green fields, over water-courses, past substantial farm houses, and barns and orchards, when suddenly the whistle blows, and we find ourselves at the charming village of

ROSELLE.
(43 min. 16 trains each way daily.)

Here the arriving passenger finds his surroundings decidedly novel and attractive. No dreary waste of dusty road stretches away, no decrepit rows of grocery stores, and liquor shops stare him into despair as he steps upon the platform and takes his first glance around him. If first impressions, as is generally conceded to be the case, are everything, the visitor to Roselle will have been prepossessed with the place before the train which brought him has disappeared in the distance. The station grounds in front and rear are laid out in lawns, flower-beds and serpentine walks, in the most approved style of landscape gardening, and are lavishly adorned with flowers and shrubbery, while a picturesque bridge, approached by a stairway at each end, spans the track, thus obviating any danger or detention from passing trains, (see cut on page 28). In short, one might easily fancy he had been set down upon some gentleman's private estate instead of at a public depot.

But come, let us cross the track. From the bridge we gain a commanding view of the road in both directions, of the adjacent landscape, and of Staten Island and Elizabeth. To the southward, not over two or three miles away, is the village of Linden, on the New Jersey railroad, and this street on which we are going runs directly thither. Here is the Mansion House, a well patronized resort in the summer season. Further to the left, and ahead, are many beautiful dwellings, including those of Chancellor Ferris (deceased) of the N. Y. University, Reuben Van Pelt, a retired merchant of New York, and several gentlemen prominently identified with the Central road and its interests. The streets are laid out at right angles, and there are good side-walks which ever way we turn. During our stroll we shall find churches of the Episcopal, Presbyterian, Methodist, and Baptist persuasions, one private and two public schools, a public Hall, in which Azure Lodge No. 129 F. and A. M. holds its communications, well stocked stores of various kinds—but—topers take notice and don't stop at Roselle—not a solitary place where liquor of any kind is retailed.

To such salutary influences as these, doubtless combined with its perfect drainage and consequent healthfulness, we may mainly ascribe the magical growth of this beautiful vil-

ROSELLE DEPOT, LOOKING EAST.

Roselle Land and Improvement Co.

HAVE FOR SALE

HOUSES,
LOTS AND VILLA SITES,

ADJACENT TO THE DEPOT,

ON EASY TERMS,

APPLY TO **A. D. HOPE,**

119 Liberty Street, New York.

OR **W. W. DILTS,**

Mansion House, Roselle, N. J.

---

lage upon a spot where but four or five years ago were only woods or open farm land. Yet, within that short time has sprung up a village of villas, with a population of about one thousand cultivated well-to-do people. Nor does it stop here. Building is constantly going on, seventeen buildings having been erected during the last half year, and there is a fair prospect that at no distant day this neat and select little borough will find itself incorporated as a ward of the great neighboring city.

Land at this point is seventy-five feet higher than at Elizabeth. Building lots may be secured at prices varying from $250 upwards, or villa plots at corresponding prices. The Roselle Land and Improvement Company offer (see above) a choice selection of property, well worthy the inspection of the intending purchaser.

Beyond Roselle our line lies parallel with the old turnpike road to Westfield, and presently brings us into the good old village of

CRANFORD,
(53 min. 17 trains each way daily.)

Cranford is a pleasant and a pretty place of about two thousand people, and has been growing rapidly of late years. The Rahway River, flowing directly through it, gives a beauty and variety to its surroundings. The streets are well laid out and kept, the sidewalks are planked from end to end of the village, and among the private residences and grounds are some that in rural beauties and elegance cannot fail to arrest the visitors attention.

There are Episcopal, Presbyterian, and Methodist churches, (the Roman Catholics will also shortly erect one) two schools, one public and one private, where the idea of juvenile Cranford is taught how to shoot, and stores where may be purchased all the necessities of daily life.

Should the visitor feel disposed to purchase here, he is respectfully informed that he could not choose a healthier locality, and that the price of lots varies between $400 and $1500.

Off again through the fertile farm lands which line our course to the ancient town of

WESTFIELD,
(59 min. 19 trains each way daily.)

which, though like Cranford founded in the olden time, has like it, begun to feel the impetus of suburban travel, and has of late attained a reputation for remarkable growth and enterprise. As we dash up to the depot it is not difficult to discover that we have reached an active, thrifty place. We see mills and spires, and business streets, and out beyond them attractive villas, just such as a city man would like to occupy with his family, while about this depot, as that at Roselle, the grounds are handsomely laid out and ornamented.

Westfield was first settled in 1720, but it was not until the completion of the Central Railroad's all rail line to New York that its progress actually began. Then, there were but 250 houses and 1500 inhabitants, where to-day there are 1000 houses and a population of over 5000, of whom two hundred are daily commuters to New York. These figures need no comment.

There are five churches in Westfield, (Episcopal, Presbyterian, Methodist, Baptist, and Roman Catholic), a well conducted and equipped public school, a public hall and library, and Masonic, Odd Fellow and Good Templar Lodges. The healthfulness of the place is proved by the longevity of its people, it being stated that in 1839 one half of the population were over seventy years of age.

Lots sell here at from $250 to $750, and land by the acre at from $2500 to $5000, according to location.

Leaving flourishing Westfield with its far stretching avenues and pleasant homes behind us, we pass through a well settled open farming country, then dash through two or three heavy earth cuttings, and emerge from the last to gain suddenly and for the first time, a glorious view of the Blue Ridge Mountains of New Jersey, distant about two miles on our right. And nestling 'neath their shadow in the intervening valley, lies the pretty village of

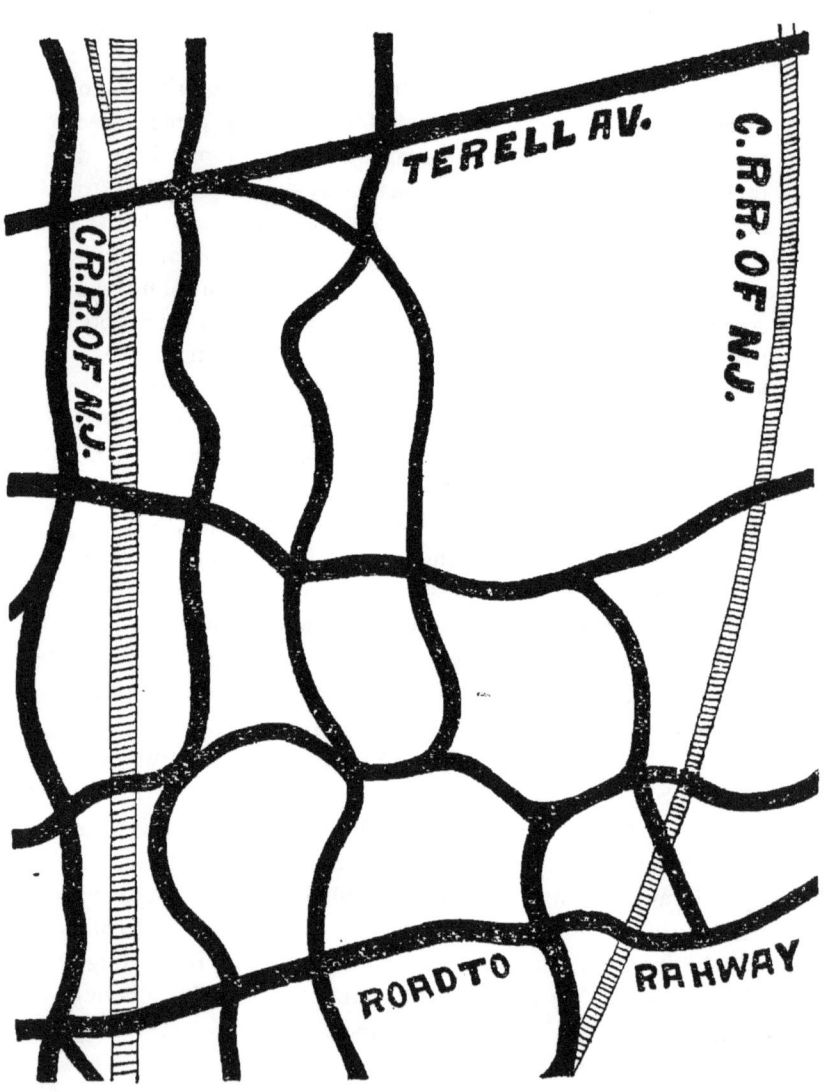

FANWOOD PARK.

FANWOOD, (OR SCOTCH PLAINS.)
(1 hour. 14 trains each way daily.)

It is only a few moments' ride to the center of the place from the station, and the agent has vehicles in waiting on the arrival of each train. The station itself is a cosy spot, shaded by oaks, and commanding a most attractive view of the back country. But we must see the village, so let us ride over. Scotch Plains, the historian tells us, was settled in 1684 by Scotch Emigrants, the spot being selected on account of its peculiar fertility. Its more modern name of Fanwood has been bestowed since the advent of the railroad, and Fanwood Park (see engraving) is now, thanks to skillful landscape engineers and its natural advantages, known as one of the most beautiful and attractive suburban dwelling places about New York. The land is gently undulating, and through it flow the waters of Green Brook, a powerful tributary of the Raritan. The adjacent mountain roads afford charming drives and scenery. In the village we shall find a population of five or six hundred, a fine public school and two churches a Methodist and Baptist, the latter of which furnished to Brown University its first President, Rev. James Manning, D. D. There are also here a public hall, a Hook and Ladder Company, a Good Templar organization, two hotels, and a variety of stores. The fine water power afforded by Green Brook has been utilized by the erection of several mills along its banks.

The spirit of improvement and development are visible here as elsewhere along the line. The re-survey of this portion of the road and other contemplated improvements by the Company will bring Fanwood Park directly upon the line, and place within a moment or two of the depot some of the most eligible villa sites that even the most fastidious purchaser could desire. Land already commands here prices varying from $500 to $2000 per acre, and is in demand at those figures. The Company will also shortly erect a new depot at this point.

Beyond Fanwood, still following the line of the mountains on our right, our course turns a little to the southwest, and presently brings us to

PLAINFIELD,

(1 l our by exp., 1 hour, 10 min. by acc. 19 trains each way daily.)

with a population of ten thousand people, and justly claiming in enterprise, convenience and beauty the foremost place among New York's suburban cities. For though Elizabeth 'tis true, has her many miles of paved streets and her wide spread improvements, Newark her railroads, and broad avenues, Paterson her mills and her beautiful Falls, Bayonne her majestic scenery and her Boulevard, Hackensack, her quaint and interesting antiquities, it may yet, without disparagement to either, be truly said of Plainfield, that proportionally she is equalled by none of them in the substantial character of her business streets, the extent and system of her public improvements, and the uniform elegance and beauty of her private dwellings and the grounds about them.

The business portion of the city, located to the right of the depot, is compactly built up with brick and stone; the remainder of the place may be better described as one vast park or flower garden, while so densely is it shaded with maples as to have won for Plainfield the not inaptly applied title of "The Maple City."

CITY HOTEL,

Cor. Second and Cherry Streets, Plainfield, N. J.

GEORGE MILLER, Proprietor.

Guests will find here all the accommodations a first-class hotel can afford.

Billiard Room attached.

GOOD STABLING.

One arriving at the depot, sees, it is true, little or nothing of these attractive features. But, if he will cross North Avenue, (which, by the way, with South Avenue, runs parallel with the railroad on either side, hence to Elizabeth), and turning to his left pass on to Cherry Street, and thence to Front, he will in a moment or two find himself in the most thickly built portion of the town. He will see the First National Bank, and the City Hotel, a model and well kept house, (see cut); the streets he will observe are lit with gas, the sidewalks paved everywhere ; some of the new brick or stone rows of stores on Front Street are Metropolitan in their size and finish. And then, if with interest excited by these evidences of thrift, he make inquiries of any intelligent bystander, he will learn that Plainfield is governed by a Mayor and eleven Councilmen, has a Fire Department, (with two steamers, one hand engine, Hook and Ladder and Hose Company) a Police Force, good sewerage, three Newspapers, two Banks, three Insurance Companies, fifteen churches, (one of which, the Second Presbyterian, is built of Ohio sand stone, and cost $75,000), a Masonic Lodge and Chapter, Young Men's Christian Association, schools, both public and

PLAINFIELD FROM PROSPECT HILL.

private of a high order, and the best of marketing facilities. If he ask how the city is supplied with water, he will learn to his surprise and delight that underlying the gravelly subsoil upon which he stands is an unfailing supply of clear, cool water, which can be obtained in any quantity and at all times by simply sinking a pipe to the distance of about twenty feet. The gravelly formation, moreover, ensuring a natural drainage renders Plainfield the healthiest of all healthy places, and entirely banishes those domestic pests, the mosquitoes, while its southerly and westerly exposures give it a genial temperature at all seasons.

PARTIES FITTING UP

HOMES ON THE CENTRAL

CAN PURCHASE THEIR

Hardware and House Furnishing Goods

—OF—

F. T. & J. VETTERLEIN,

(SIGN OF THE BIG PADLOCK,)

Front Street. near Somerset, Plainfield, N. J,
AT NEW YORK PRICES!

And now, having inspected the heart of Plainfield, let us wander out toward its extremities. Its pleasant shady streets stretch away in all directions. First, we will take a look at North Plainfield. Just over Green Brook, which passes close to Front Street, and divides Union from Somerset County, our walk brings us in full view of the mountain again, the slopes of which already indicate that they are to be soon occupied by stately villas. A stone-paved roadway from the city to the mountain is, in fact, being already constructed.

About a mile distant are the picturesque Wetumpka Falls. Now we turn down Grove Street, and see before us square after square filled with tasteful dwellings, all of them fitted up as conveniently as city houses. The Washington Park grounds, comprising about three hundred acres, and imperatively restricted against nuisances, are located just beyond, and driving through the serpentine roadways, one knows not whether most to admire the scenery of valley and mountain beyond, or the taste displayed in the buildings and grounds before him. Yet three years ago this was all an open farm.

Then, crossing Green Brook again, we may pass through Center Street, and, after riding past a succession of shaded, smooth cut lawns, and cozy homes, most of them surrounded with broad verandahs, may reach Prospect Hill, where we obtain the capital view herewith presented of the city, half hidden among the luxuriant foliage.

But, by a paradox, what must be termed the West End of Plainfield is that section of the city lying to the east of its business center. Here the visitor will find among many other charming places of residence, those of John Taylor Johnston, Esq , (a view of which is presented herewith), and of the son of the lamented Admiral Farragut.

With such natural attractions and advantages as these the growth of Plainfield to the dimensions of a large inland city is simply a question of a very few years. Its growth and development within the past nine years have been so rapid and substantial as to justify the great expectations of its people with regard to what is in store for it.. The proposed erection of a new and elegant depot by the Central R. R. Company, and the depression of the city streets beneath their track ; the liberal inducements held out by real estate owners;

RESIDENCE OF JOHN TAYLOR JOHNSTON, ESQ., PLAINFIELD.

GEO. A. MARSH,
Real Estate Agent,
OFFICE OPPOSITE DEPOT,
P. O. BOX 671,   PLAINFIELD, N. J.

and, perhaps more important than all, its perfect healthfulness, all combine to mark out for Plainfield a brilliant and prosperous future.

The purchaser can secure here good building lots (50 x 100 feet) at from $1,000 to $1,500, and can build at a cost of $3,500 and upward, according to the size of his family and his purse.

It is only after leaving the Plainfield depot that the passenger, who has not alighted, gains any correct estimate of the extent and beauty of the city. Street after street stretches away on either side, then finally the buildings grow fewer, and we are once more in the open country. Only for a few moments, however, for here we are at

EVONA,
(1 hour and 12 minutes. 6 trains each way daily,)

Where, as if by magic, have recently sprung up, upon the greensward, as Robin Hood's "merry men" were wont of yore to spring forth when a rich bishop's train was passing through their domain, houses and stores and all the characteristics of a thriving suburban village. The depot at this point is a remarkably large and handsome one, and is surrounded by a park laid out on a liberal and tasteful scale.

Land near the depot sells at from one to three thousand dollars per acre.

Evona boasts a population of about three hundred, and relies on Dunellen, only three-fourths of a mile distant, for her church and school facilities. In fact, even now we are virtually within the limits of

DUNELLEN,

(1 hour and 14 minutes. 14 trains each way daily,)

Dunellen is charmingly located in full view of the whole valley and the mountains beyond, and is moreover a growing and thrifty place, as the view on page 42 indicates. Like many of its sister stations with romantic names, it owes its origin to the completion of the Central's all rail route to New York in 1865, and the consequent demand for country homes for business men. But it has been and is steadily growing, and has now several hundred inhabitants, two churches, schools, stores, a good market and a hotel. There is fair water power offered manufacturers on the streams back of the village, and the sportsman may be sure of good hunting and fishing in the immediate vicinity.

Land sells here in plots of 50x100 or 150 feet near the depot for from $6 to $20 per foot, and that more remote, say half a mile away, at from $2000 to $3000 per acre.

A mile and a half south is the beautiful village of Newmarket, by which name in fact the station was known until called by that it now bears.

Dunellen derives, moreover, an additional importance as the nearest connecting point for Washington's Rock, a bold cliff four hundred feet high, plainly visible on the face of the adjacent mountain about a mile distant, and from the summit

42   HOMES ON THE CENTRAL.

DUNELLEN.

HOMES ON THE CENTRAL.

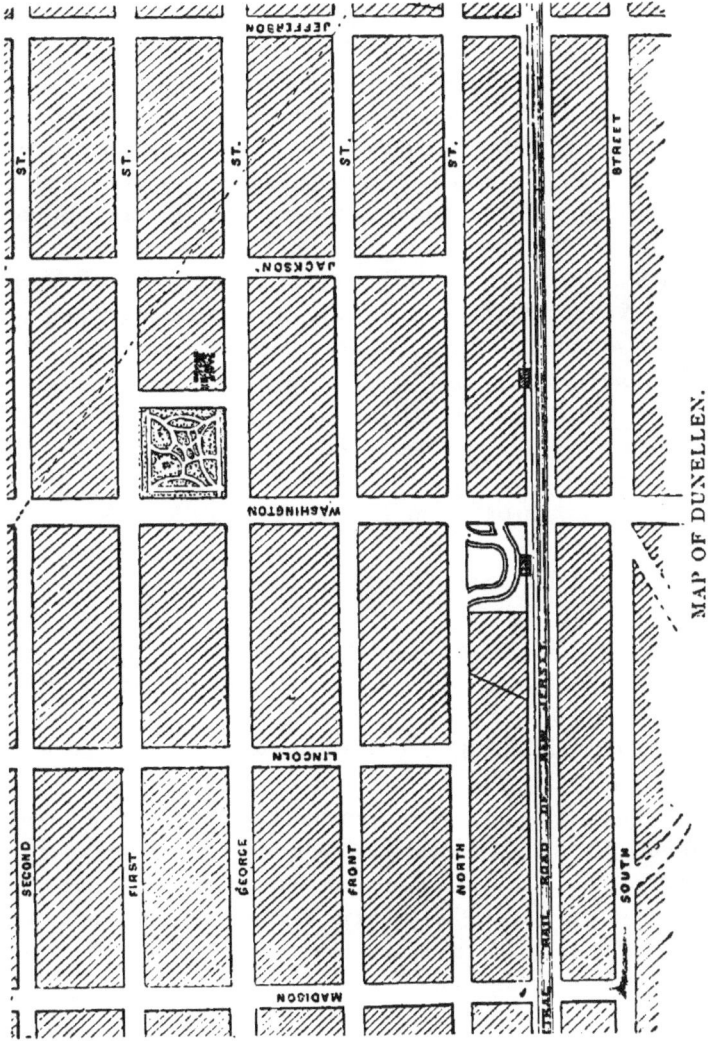

MAP OF DUNELLEN.

of which the revered patriot whose name it bears, was wont, during the campaign of 1777, to watch the movements of the enemy. During the skirmish between the troops of Sir Wm. Howe and Lord Sterling, near Plainfield, Washington was on this rock inspecting the movements of the two armies on the plains below. For many years past this spot has been a favorite one of resort for pleasure excursion parties, not for its historical associations alone, but for its majestic view which embraces an area of sixty miles, including New York City, Newark, Staten Island, Raritan Bay, the Highlands of Navesink, New Brunswick, and the heights of Princeton and Trenton. In short, one sees mapped out before him, and dotted with countless villages, towns and cities, the entire stretch of New Jersey landscape, from the Hudson to the Delaware.

Excursion parties from New York, Newark or Elizabeth can conveniently reach the Rock early in the forenoon, enjoy a delightful day amid its surrounding beauties, and return home before dark.

In a spot so enchanting, it were tempting to linger longer. But our iron horse snorts, impatient for the many miles yet before him, so let us be off again. Another moment brings us to BROOKSIDE, formerly known as West Dunellen, and an embryo suburb of the prosperous parent town. Deriving its name from its proximity to the beautiful Green Brook, which passes directly through it, Brookside presents to the seeker for a rural home, many intrinsic attractions. It fronts directly upon the Central Railroad, and is intersected by the main avenues of the county, connecting with both Plainfield and Somerville.

HOMES ON THE CENTRAL. 45

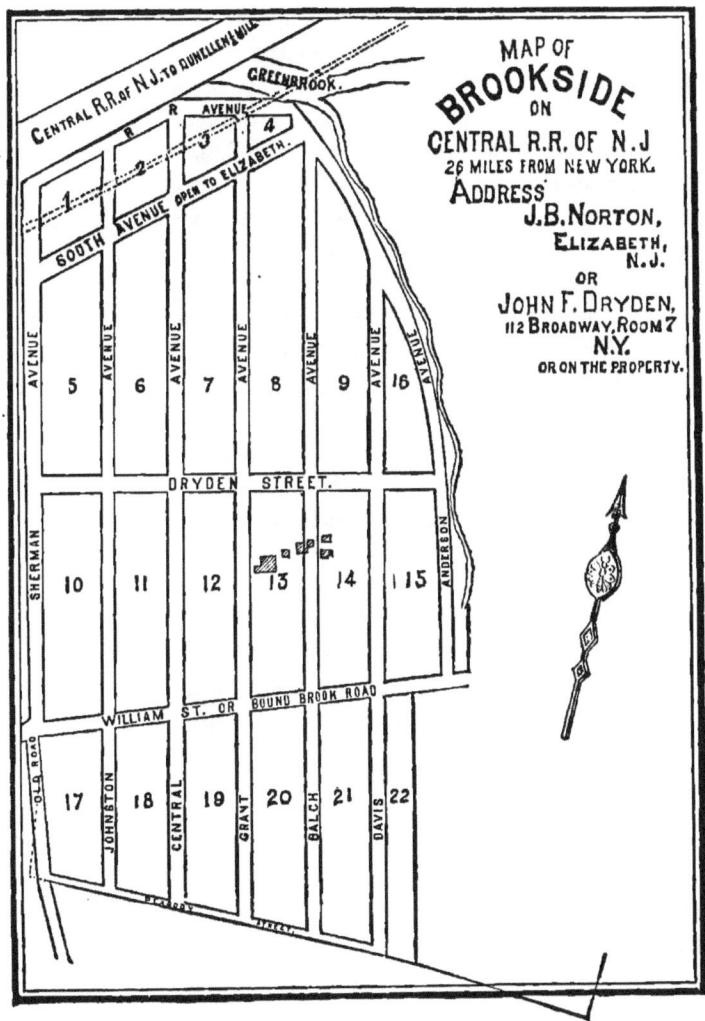

Here is a site upon which it is hoped at no distant day will spring up a thriving and attractive village, laying claim to the attention of New York business men. The scenery and surroundings are especially charming. The historical Washington Rock is here, as at Dunellen, in plain view on the adjacent mountain side, while hill, valley and meadow-land scenery, fine drives and excellent roads, combine to make a residence here one of undoubted attractiveness. Within a mile of the place are no less than five churches, and the excellent school and store facilities of Dunellen, while, for healthfulness, one could not find a spot more free from malarial or contagious diseases, there being no low swampy ground in this section of the country.

The diagram which we publish herewith directs the purchasers attention to the improvements which have been made at this point, and conveys a fair idea of its convenience and advantages. Building sites on high well drained ground may be secured at most reasonable terms, or at prices varying from $90 to $500 per lot. In view of its promised prospective growth and improvement, property at Brookside may prove a safe and profitable investment.

Our next stopping place is the time honored village of

BOUND BROOK,

(1 hour, 22 min. 15 trains each way daily.)

named from a neighboring water course forming the boundary line between Somerset and Middlesex Counties. Others attribute its title to the fact of its being bounded by brooks or rivers on every side. The former derivation appears however to be the best substantiated of the two.

Bound Brook is mentioned in Smith's History published in 1765, as being a village. In the winter of 1778-9, a portion of Washington's army were in barracks in the vicinity. The place has always been a thriving one, its location on the Raritan River and more recently the Raritan Canal, rendering it quite an important shipping point, especially in grain. There has also been more or less manufacturing carried on. The view from the car window on the left as we stop at the depot, embraces what was formerly South Bound Brook, now Bloomington The elevated and beautiful lay of the land, and the elegant residences, with masts of shipping in the foreground, form an exceedingly pleasant and rather novel scene at an inland town.

E. VAN SYCKEL,
Land Purchasing Agent
BOUND BROOK, N. J.
Investors should Apply Early.

The Delaware and Raritan Canal of 9 feet water, has locks of 210 feet in length and 25 feet in width, and forms part of the great inland water thoroughfare for Steamers, Barges, Propellers, Schooners and canal boats of every description from New York and the East to Philadelphia, Baltimore and the South, besides delivering the tonnage from canals running into the coal fields of Pennsylvania. Upon its banks are seen rare facilities for manufacturing and for general business.

Bound Brook is the diverging point in roads running south or west from New York, and bids fair to become an important railroad center, because of the protection on the north by the mountains, and on the south by the canal-draws, which are only to be avoided by running new roads through this place.

J. W. PRATT'S

STEAM

Book and Job Printing

ESTABLISHMENT,

75 FULTON STREET,

NEW YORK.

A Boulevard of scarce two miles length, from the river to to the mountain slope, is under agitation as among the requirements of the place. It should be of sufficient width to contain a horse car railroad through its center, connecting the depots on all the intersecting railroads. This straight shoot from the river to the mountain will show the great necessity as well as the proper location for the new and elegant iron bridge so long talked of for spanning the river at this place. The mountain slopes at the rear of Bound Brook, like the heights of Bloomington at the front, afford the most charming landscape scenery; in the distance are to be seen the church spires of Plainfield, Dunellen, New Brunswick, Middlebush, Millstone and Somerville, and equally as charming are their bells heard on the quiet Sabbath morning calling to the sanctuary.

The population of Bound Brook and Bloomington are about equal, and together full 1200. These places contain five churches, five schools, a Masonic Lodge, three hotels, and extensive lumber yards, where those contemplating the erection of Homes on the Central can purchase building material at reasonable figures, and have it shipped direct. (See advertisement of L. D. Cook & Co.) There are a variety of stores and shops to supply the demands of daily life. This valley of the Raritan abounds in fine drives, fertile lands and general healthfulness. The scenery from the mountain top is grand, extending as far as the eye can reach, and is unsurpassed in its beauty.

L. D. COOK & CO.,
LUMBER DEALERS

Near the Rail Road Depot, Bound Brook.

Constantly on hand, a full assortment of all kinds of Lumber, planed and in the rough. Scranton and Lehigh Coal, at the Lowes. Prices.

Frames for Buildings cut to order at Short Notice, and delivered at any point on Central Railroad, N. J. and Delaware and Raritan Canal.

LEWIS D. COOK.   G. R. GILES.

Improved farms a mile or two from the depot can be obtained at about $200 per acre, but the most convenient lands are priced as building lots at much higher rates.

About a mile and a half from the village, the sight-seer will find in a wild and romantic ravine the famous "Chimney Rock," a singular pyramidal shaped stone, fifteen or twenty feet high, on the summit of a bold and nearly perpendicular ledge over one hundred feet high. Tradition says that an Indian pushed his wife off this rock.

Close at hand, too, is another attractive resort, Buttermilk Falls, where a few years ago Blondin performed one of his dangerous exploits on the tight rope. Had the Indian Blue Beard lived at the present day, it is possible that, under Jersey justice, he, too, might have become famous by a tight rope.

But here we go again. Now we cross Middle Brook by an iron bridge, catch a view of the level landscape stretching far away before us to the distant blue mountains in the south-west, and presently stop again at

FINDERNE,
(1 hour, 27 min. 9 trains each way daily.)

a small station with, however, a quite handsome depot. There is here a population of not over an hundred, with a school. The place is in reality a sort of suburb of the important county town which we are now approaching.

SOMERVILLE
(1 hour, 24 min. by exp., 1 hour, 34 min. by acc. 15 trains each way daily.)

is pleasantly situated on a knoll on the right of the railroad, and directly on the line of the old turnpike road from Elizabeth to Easton, which, passing through it, forms its main or principal business street. The village itself is comparatively of modern date. During the revolution a tavern was kept on the site of the Somerville House, but it was not until the burning of the Court House at Millstone by the British in October, 1779, that this was made the county seat. In 1784 a log court house and jail were built here, the former about twelve rods east of the present court house, which was erected in 1798. A visit to the court house, and a view from its cupola, should be the visitor's first object on arriving at Somerville. Let him pass up Railroad Street from the Depot to Main Street, then turn to his right, and a few steps will bring him to the spot where, in the midst of a handsome park and embowered in a dense growth of foliage, stand the county buildings, overlooking the village and all the adjacent country. Having admired the view from this point, let the visitor retrace his steps, and stroll along the shaded sidewalks of Main Street to and through the western portion of the village. He will find churches of the Dutch Reformed, Methodist, Presbyterian and Baptist denominations, Masonic K. of P., Good Templars, and Sons of Temperance Lodges, four hotels, one large public school and several private ones, two banks, a savings bank, and no less than three newspaper offices. The beauty and taste displayed in most of the private residences will surprise and please him, while what will especially claim his admiring attention is the number and luxuriance of the trees, which provident hands in days gone by have planted along the streets, and in almost every door yard and lawn, to shade and beautify the homes of generations to come after.

The importance of shade trees in laying out one's country home can not be too strongly urged. How many associations of a childhood's home are interwoven with the memories of the old trees which waved their branches over its door. How often on the dusty road of life's noonday does the traveler look back to recall the grateful shadows of the boughs though which his eye "first looked in love to the summer sky." And to how many a wanderer far away from his native land, has the home tree been the Mecca to which all his hopes and longings have been anchored.

> "Yes! when thy heart in its pride would stray
> From the pure first loves of its youth away—
> When the sullying breath of the world would come
> O'er the flowers it brought from its childhood's home—
> Think thou again of the woody glade,
> And the sound by the nestling ivy made,
> Think of the tree at thy father's door,
> And the kindly spell shall have power once more."

Somerville has a population of about three thousand, and is growing at a moderate rate. Among its projected improvements is a horse railroad connecting it with Raritan and Bound Brook. In common with Plainfield, and its other sister towns, it enjoys a healthful atmosphere, while its quiet beauty gives it a charm to those in search of a peaceful retreat from the din and dust of the city.

Land sells in the village at from $150 per lot upwards, while, within a mile or two from the depot, it can be purchased at $300 per acre.

From Somerville the branch road to Flemington diverges. That thread we shall follow up on some other day. For the present our line lies straight on to

RARITAN,

(1 hour, 39 min. 4 trains each way daily.)

a lively manufacturing village of twenty-five hundred inhabitants, situated on the north bank of the Raritan, which here, with a fall of 16 feet, furnishes motive power for extensive woolen mills, and agricultural machine shops. There are here three churches, (Dutch Reformed, Methodist and Roman Catholic), a public and a private school, a savings bank, good stores, and a market. Public enterprise has also displayed itself in the organization of two Building Loan Associations. Land sells in the village at $50 per foot on the main street, and from $6 to $20 per foot on the side streets.

Good farms a mile or two distant can be bought at $200 per acre.

Among the residences on the river bank, near the village, is that of New Jersey's distinguished statesman, Senator Frelinghuysen.

NORTH BRANCH,
(1 hour, 46 min. 7 trains each way daily.)

our next stopping place is the point at which passengers alight for the village of the same name, distant about one mile to the northward, and situated, as its name indicates, on the North branch of the Raritan River. It is on the old Somerville and Easton turnpike, and has a population of about six hundred, with a Dutch Reformed church, a school and three stores.

The growing popularity of this immediate locality as a place of residence for men who have to some extent withdrawn from active participation in business affairs, yet find it necessary to visit the city occasionally, is worthy of notice. Here may be found many attractive sites upon which such purchasers may locate within half an hour's ride from the depot, and enjoy, in all its perfection, country life within a short distance of the city. Let us, for instance, visit Hopewood, a model farm of about seventy-five acres, lying on a verdue-clad knoll to the south of the station. Our ride thither brings us along the wooded banks of the north branch of the Raritan, and opens many a lovely sylvan vista, at one time seen upon the level with our road, at another looked down upon from overhanging bluffs and through luxuriant foliage. But presently we turn to the right, leave the river behind, and by an ascent almost imperceptible find ourselves entering the gateway and traversing the serpentine roadway through the lawn at Hopewood. But how remarkable—we are on but a slight elevation, yet, we command a complete view of the horizon for fifteen or twenty miles distant, on all sides, can see Plainfield, New Brunswick, Whitehouse—well fifteen important towns and villages in all, and a landscape so fair and cultivated as to remind one of Longfellow's description of peaceful Acadia.

Before us is the house, large, modern and commodious. Behind it are barns and outhouses, on every side orchards of and peach and apple trees, fields of corn and wheat and oats stretch away, enclosed by a hedge of evergreens. The peculiar fertility of this soil should alone suffice to recommend it to the purchaser. Grapes grow profusely, as indeed do all kinds of fruit and grain, and abundant harvests of any kind reward the tiller.

Just beyond North Branch, we cross Chambers Brook, and, pass through a beautifully diversified region of hill and dale; here we see great knolls rising almost abrutly from the landscape, yet cultivated to their summits, and forming in their vari-colored grain fields a natural patchwork or mosaic. Then we see farm houses and pasture grounds and presently come to the village of

WHITE HOUSE.

(1 hour, 56 min. 8 trains each way daily.)

For a time, in the early history of our road, this point was its terminus, from which passengers were booked through by stage to Easton and Delaware Water Gap. Then it was a mere hamlet, but later years of railroad communication have developed it to a village of considerable size. It derives its name from an old settlement about a mile to the north, on Rockaway Creek and on the turnpike road before mentioned, where stands an ancient dwelling at which Washington is said to have halted to take dinner.

The present White House has about seven hundred inhabitants, Dutch Reformed and Methodist churches, two schools and three hotels. About a mile to the south is the village bearing the euphonious name of Scrabbletown.

Farm land can be purchased hereabouts at $80 or $90 per acre. Quarter acre lots in the village sell at from $200 to $400.

It will be observed that since leaving Somerville we have passed through a region which, though comparatively remote from the Metropolis for men engaged in daily active business pursuits, is yet rich in attractions as a place of residence for those who wishing the quiet and repose of a rural home, are yet desirous of visiting the city two or three times a week, or even for an hour or two daily. One may leave White House for instance, about half past ten, pass three or four hours in town, and yet be home again in season for an early supper, or before six o'clock. For those who have passed the hurry and bustle of life, retired business men, and men of studious habits, who would live within easy distance of New York, yet be in the midst of perfect rural repose, no section of New Jersey can be more enchanting than this picturesque and fertile valley of the upper Raritan.

And now, as we leave White House we see confronting us on our left, and thence stretching southward, the rugged slopes of the Pickles Mountain, and to the right, broken into gentle undulations, a well tilled farming country. Presently we enter the Lebanon Valley, where to the right we discern, standing boldly out against the hills in the background, the little village of

LEBANON,
<small>(2 hours. 6 trains each way daily.)</small>

situated in the center of a fertile tract, literally flowing with milk and honey. Nearly one hundred cans of milk are shipped hence to New York daily, and a large creamery is among the important industries of the place. Many fine peach orchards adjoin the village, which itself has a population of about three hundred, with Dutch Reformed and Methodist churches, an Academy, a High School, Hotel and Post Office.

The price of land varies from $150 to $500 per acre.

After leaving Lebanon our line intersects that of the turnpike road to Easton, and presently brings us to

ANNANDALE,
<small>(1 hour, 54 min. by exp., 2 hours, 6 min. by acc. 8 trains each way daily.)</small>

formerly known as Clinton, and earlier still as Hunt's Mills, a Mr. Hunt having been one of the early proprietors of the valuable water power furnished by the South Branch of the Raritan at this point. In 1820 there were but three houses here. A Post Office was first established in 1838. The Presbyterian church was erected in 1830, the Episcopal in 1838, and the Methodist in 1840. The village proper is distant about one mile from the depot, and still retains its post office name of Clinton, that at the railroad being known as Annandale. Stages run to and fro on the arrival of every train.

The adjacent region is very fertile, and as many as 5000

baskets of peaches have been shipped in a single day. There are also numerous beds of limestone in the valley, while in the mountains, mines of hematite and magnetic ore exist, and are about being developed by a company recently organized for that purpose.

For manufacturers, this spot offers undoubted advantages. Land sells at $200 per acre, or town lots (50 x 175) at $500. At

HIGH BRIDGE

(2 hours, 11 min. 6 trains each way daily.)

we cross the Raritan's south branch and valley by an embankment (formerly a wooden bridge only), thirteen hundred feet long, and one hundred and five high. The view both north and south, as we are whirled over this great viaduct, is superb. We look down upon roofs, tree tops, the river, and a mosaic of cultivated fields far below us, while stretching away in the distance are a succession of fertile uplands terminating in distant blue mountains.

High Bridge derives its existence, as it does its name, from

TAYLOR IRON WORKS,
HIGH BRIDGE, N. J.

New York Office, -   -   -   -   93 Liberty Street.

MANUFACTURERS OF

Car Wheels and Car Axles.

LEWIS H TAYLOR, Pres't. JAS. H. WALKER, Sec. & Ass't Treas.
W. J. TAYLOR, Treas. & Man. S. P. RABER, Superintendent.

E. L. BROWN, General Agent.

the railroad. Yet, within the past few years it has attained a population of twelve hundred, and now presents, with its three churches, Dutch Reformed, Methodist and Roman Catholic, its 3 hotels, its Iron Works, (see advertisem'nt and views), employing 150 hands, and its numerous large stores and dwellings, quite an imposing appearance to the visitor.

Village property is quoted here at from $500 to $1000 per acre, 6 lots. Farm lands can be had at $100 per acre.

TAYLOR IRON WORKS—FORGE.

TAYLOR IRON WORKS—CAR WHEEL, FOUNDRY AND FITTING SHOP.

These Works possess peculiar advantages, having an unlimited water power, easy communication in every direction, and proximity to the coal fields of Pennsylvania, making both wheels and axles, and having ample facilities for "fitting." The Company can supply railroads and car builders with wheels and axles ready to go under cars at but short notice.

Our route, since leaving Annandale, has been toward the northwest. And now presently we skirt the mountain side on our left, and on our right look down upon the Easton turnpike again, here lined by a dense growth of evergreens, through which flows the water course known as Spruce Run. The scenery here is wilder than any we have hitherto seen on our route, and forms an appropiate setting for the exquisite picture which awaits us on our arrival at

GLEN GARDNER,
(2 hours, 20 min. 6 trains each way daily.)

situated in a romantic and picturesque dell among the mountains, and presenting in its busy streets, its substantial buildings, its handsome churches and dwellings, and its extensive manufactories, all the characteristics of a thrifty, growing community.

In its younger days it was known as Clarksville, a title for which that more romantic one which it now bears was a few

years ago substituted. If we alight, and first stroll up the Glen we shall be surprised and delighted at the industries

Gardner M'f'g Co.
Factories, Glen Gardner, N. J.

WAREROOMS, 110 BOWERY, NEW YORK.

MANUFACTURERS OF

CHAIRS.
Also, Chair and Car Seats for the Trade.

---

and improvements which the enterprise of man have wrought. Prominent among these are the extensive chair works of the Gardner Manufacturing Company, a view of which we give the reader herewith. These works employ nearly two hundred hands, and not only contribute largely to the life and activity of the village, but supply the lower part of it, or that part lying on the left of the railroad, with gas and water.

If after inspecting the works, we turn our steps to the valley below, we shall see there, too, abundant evidence of public spirit. There are in all, three churches, (Methodist, Presbyterian and Lutheran), two schools, a hotel, a large public hall, and Masonic, Good Templars, S. of T., and O. U. A. M. Lodges.

Nor are the neighboring mountains lacking in productiveness, for beneath their rugged slopes are to be found rich veins of iron, which are now being worked.

The purchaser who would establish himself as a manufacturer or resident in this bustling little place, with its population of fifteen hundred, can secure building lots (50x100) at from $250 to $1000. Good farm land over the hill sells at $200 or $300 per acre.

Five minutes further ride through some heavy cuttings in the mountain side brings us to

NEW HAMPTON JUNCTION.

(2 hours, 10 min. by exp. 8 trains each way daily.)

Here, we see, on our right, the diverging broad gauge track of the Del., Lack. and Western Railroad, and here passengers for Washington, Belvidere, Delaware Water Gap, and the many beautiful spots in the north and east portions of Warren County, change cars. Directly under the shadow of the neighboring mountain, the base of which the road reaches by a long and almost dizzy descent along its side, the tourist will find one of the loveliest of sylvan spots, the little hamlet of Changewater, where the Musconetcong River flows through shaded banks, and under the shadow of an old stone mill and bridge.

But we must not wander off too far from our main line. Quite a settlement has grown up about the junction, but the principal village of New Hampton is on the river in the valley below. It was called New Hampton half a century ago. The visitor will find it a quaint rambling place, with a population of about a thousand, several churches, a hotel or two, and many cozy homes.

And now, resuming our journey, our road makes a sharp turn to the southwest, and we find ourselves skirting the slope of the Musconetcong Mountains overlooking the lovely valley named from the river of the same name which winds its way at their base. Certainly in all New Jersey there can be found no more fertile or peaceful landscape than that which, stretching away to the Pohatcong range, lies mapped out below us. Let the traveler by all means secure a seat on the right hand side of the car during this portion of the ride. He will have an opportunity to enjoy a panorama of vivid and ever changing beauty; a landscape completely cultivated, and dotted with villages and spires and clusters of farm houses. Under the shadows of the opposite range, the Morris and Essex railroad locomotives push their rapid way, and the horses drag their weary loads along through the sluggish waters of the Morris Canal. The valley gradually narrows as we advance, and very soon the whistle sounds for our stop at

ASBURY.

(2 hours, 34 min. 7 trains each way daily.)

The village of the same name we can distinctly see nestling in the valley about a mile distant. It is an old place, and was once called Hall's Mills, but in 1800 the corner stone of the old Methodist church was laid by the venerable Bishop Asbury, in whose honor the village was re-named.

There is a mine in the mountain at this point, and, in the township, a mineral spring said to be nearly equal to that at Schooley's Mountain.

Asbury has a population of four hundred, two churches, a school, a hotel, three large grist mills, and a basket factory.

VALLEY,

(2 hours, 38 min. 5 trains each way daily.)

our next stopping place is the connecting point for the village of Bethlehem, about half a mile to the south. It has a school, church, (Methodist), hotel, and a population of about two hundred. At

BLOOMSBURY

(2 hours, 45 min. 8 trains each way daily.)

we cross the Musconetcong, now grown to a stream of considerable dimensions. From the summit of the neighboring mountains may be obtained a splendid view, stretching over a great extent of country, and even comprehending the distant city of Easton.

The principal part of Bloomsbury lies on the south side of the river. It has Presbyterian and Methodist churches, a good school, a hotel, four stores, and a population of seven hundred. At

SPRINGTOWN
(2 hours, 50 min.   4 trains each way daily.)

our route crosses the Pohatcong Creek. The village which is about a quarter of a mile southwest of the depot, has a church, school, hotel, and only about one hundred inhabitants.

And now our course turns once more to the northwestward, and presently brings us to

GREENWICH,
(2 hours, 55 min.   3 trains each way daily.)

where once more, and for the third time on our journey, we cross the Morris Canal. Yet we have accomplished in three hours the distance which, by its tiresome towpath, requires a journey of at least as many days. The view at this point is extremely picturesque and attractive. The village itself is a small one, deriving its name from the township, and contains one church and a school.

And now we near near the Delaware river, and the western border of the noble little Commonwealth which gives our railroad its name. On all sides of us as we advance, we discern indications of our approach to another great business center, and presently see on our left the smoky chimneys and forges of the Pittsburg of New Jersey.

PHILLIPSBURG,

(3 hours. 8 trains each way daily.)

The visitor, fond of mechanical industries, will find in a stroll through the workshops of Phillipsburg fully as much to admire as he will in the beautiful scenery that surrounds it. But a view even more striking and beautiful awaits him when he resumes his ride. For, in a moment after leaving the depot, he finds himself whirling along as it were in mid-air with the Delaware far below him, and side by side with the immense bridge over which he is passing, is another equally imposing structure. And now, if he look to the southward, he sees the river winding and finally losing itself from sight amid bold high banks, where the hand of man, as if in defiance of nature's obstacles, has planted countless dwellings and manufactories. To the north, however, he turns to catch the brightest and most inspiring side of the picture. On the right, the upper part of Phillipsburg overlooks the stream, a densely built locality, teeming with life and industry. To the left, the Lehigh's waters commingle with the Delaware's and, ensconced beneath the hills which form the background of the charming picture, lies the active and prosperous city of

EASTON.

(3 hours. 3 min. 8 trains each way daily.)

To the north of the city, the eye rests upon the spacious and elegant buildings of Lafayette College, while a glance over the intervening space reveals a crowded medley of spires, and domes, and chimneys, extending back from the river as far as the eye can reach. Now, we are over the Delaware, cross a bold promontory of solid rock, through which the skill of the engineer has hewn for us a cutting fully an hundred feet in width, and—but what is this? another bridge? yes, this is the Lehigh we are crossing now, and the view is totally changed. Glancing up its current, we see South Easton on the southern bank, with numberless foundries beyond, the river itself dotted with canal boats and smaller craft; and to the right, and ahead and beneath, the city which is our temporary destination.

No one who has visited Easton can truthfully deny that it is an unusually neat and pretty city. When we alight and stroll or take the horse car up South Fourth Street, and thence to "the circle," the name given to the public square in the heart of the city, we shall not fail to be pleased with the cleanliness of the streets, the abundance of well grown shade trees, and the substantial character of the private dwellings, principly built of brick or free stone. One sees much in their style and finish to remind him of the Quaker City.

Then, in the business portion of the city, one sees in the rows of handsome stores, the banks and the large new Opera House, gratifying evidences that the enterprise which brought hither the early German settlers who founded the city is not lacking in their descendants. Easton is, in short, the seat of wealth, culture and refinement, and the New Yorker strolling through its delightful streets, or experiencing its courtly hospitalities has but one regret, "What is it," does the reader inquire? Why, that it is not an hour or two nearer the Metropolis, so that he might daily repair from the toil of business to find repose amid its grateful shadows.

FROM EASTON TO MAUCH-CHUNK

Our line skirts the Lehigh River, passing through one continuous and varied scene of beauty. We pass furnaces and foundries innumerable, lining the river bank, and possessing in the aggregate an annual capacity of hundreds of thousands of tons. Thus we are whirled in turn to GLENDON and FREEMANSBURG, and reach our first important stopping place, the quaint old Moravian mission city of BETHLEHEM, which the latter days of railroad enterprise have transformed into a thriving and prosperous commercial center. Here the tourist will find many objects to interest him. The immense works of the Bethlehem Iron Company and the Lehigh Zinc Company; the spacious and beautiful buildings of Lehigh University, founded, and thrown open *free of expense* to the youth of the country by the munificence of Hon. Asa Packer of Mauch-Chunk; the view from Nisky Hill, where is situated one of the most tastefuly laid out cemeteries in the State—all these will in turn attract and interest the visitor at Bethlehem.

At Bethlehem we intersect the great line of pleasure travel from Philadelphia to Saratoga, Sharon Springs, Albany, Lake George and Canada. This route, newly opened, carries the traveler through the picturesque coal regions of North-eastern Pennsylvania, and up the fair valley of the Susquehanna, affording a most convenient and attractive route for tourists and pleasure-seekers during the present season.

Only five miles further on we came to the busy city of

ALLENTOWN, with its 16000 people, situated in the midst of a rich agricultural district, in close proximity to valuable beds of iron ore, zinc, limestone and cement, and connected by rail with all four points of the compass. Here are manufactured iron and steel rails, engines, machinery, carriages, fire-brick, and a host of lesser useful articles. Here, too, are two educational institutions, Muhlenberg College and the Allentown Female College. Here are the Fair Grounds, annually visited by forty thousand people; here a bridge of 19 arches, 1800 feet long and 59 high, spanning Jordan Creek, and here the visitor will find scenery and natural curiosities well worthy a day's leisure. The view from Big or Bauer's Rock, a thousand feet high, embraces a rich variety of landscape in the Lehigh and Saucon valleys, and there are several romantic springs much frequented by summer visitors.

CATASAUQUA, four miles further up the river, is the point where anthracite iron was first successfully manufactured in the Lehigh Valley. Now 25000 car wheels alone are annually made here. Just above the town stands a building nearly two hundred years old, once occupied by George Taylor, one of the signers of the Declaration of Independence.

And now we pass in turn LAUBACH, SIEGFRIED BRIDGE, TREICHLER and WALNUT PORT, and, in little over an hour after leaving Easton, find ourselves amid the grand scenery of the Lehigh Gap, where the river forces its way through the Blue Ridge. To many the beauties of this point would prove a sufficient temptation to alight. But the knowing one will tell you that there are even grander beauties beyond, so let us ride a few miles further. Now, we see busy PARRYVILLE with its three furnaces, and then WEISSPORT and LEHIGHTON situated on the opposite side of the river. Here we cross by a substantial bridge from which we gain a novel and striking view both up and down stream. As we pause for a moment at the depot, we may be interested in knowing that this locality has some interesting traditions of its own. It was first settled in 1746 by Moravian Missionaries, eleven of whom in 1755 were murdered by Indians from Canada. The mission house was burned at the same time.

In Weissport is the site of a log hut built by Benjamin Franklin when he was in charge of the then northwestern frontier.

Beyond Lehighton we have the river on our right, and the hills gradually closing in more closely about us. What a pigmy our train seems to be, winding its way through the verdure clad battlements towering above it. Here we see PACKERTON with its park of seventy-five acres stocked with antelope, deer, elk and trout, and presently a sudden turn in the valley brings us suddenly in view of our destination

, MAUCH CHUNK,

nestling lovingly in the embrace of the giant mountains about it.

Alighting for the first time in this picturesque spot, where the enterprise of man has engirdled with railroads and canals, the wildest mountain solitude, and has brought within a half days pleasure ride of the Metropolis, regions once deemed all but inaccessible, one knows not whether first to bow in awe at Nature's majesty, or exclaim with delight at the triumph which engineering skill has achieved in bringing it so readily within our grasp. For see! this narrow gorge through which the Lehigh through ages of solitude plashed its way seaward, now furnishes an avenue for two railroads, a canal, and, at this point, for a village street, all crowded into this narrow space and monopolizing every inch of room they can ever possibly expect to occupy,

Look out of the window before we alight. That cone-like mountain opposite is Bear Mountain, or in Indian language "Mauch Chunk." Under its shadow is the brick depot of the Lehigh Valley Railroad, and from its platform a bridge spans the river just before us. Then, looking this side of the depot, we see that it stands upon a walled embankment, below which winds the Lehigh Canal; then comes another walled embankment, and below it, upon the third, or lowest level, the river flows over its rocky bed. On this side, our railroad runs on the level of the street, while facing us, and, built almost into the mountain side, stands that famous hotel the Mansion House.

Fortunately it is our destination, and the cars have brought us from Jersey City to its front door. These wide verandahs with their cozy arm chairs, and their view of everybody and everything, were sufficient to tempt one twice the distance. So, without unnecessary delay, we hasten to our rooms, repair our toilets, take a royal dinner, light a cigar, and prepare to do Mauch Chunk.

Now, first make up your mind to do it coolly. Don't rush about in a hurry, and fancy you are leaving something unseen. In that way you will enjoy nothing. A day will suffice to see the whole place comfortably, provided you go about it systematically.

First, we will sit down here on the hotel piazza, and take

MAUCH CHUNK FROM THE SOUTH.

our bearings. This gigantic mountain south of us, which, turning abruptly to the eastward, seems at first glance to offer an impenetrable barrier to the river's progress, is known as the Flagstaff, and along its rugged slope can be discerned a carriage road which, engineering skill has constructed as another means of communication with Lehighton and Packerton. A rustic foot path, within a few steps of where we sit, points the way up this steep mountain side to a famous ledge, entitled "Prospect Rock," commanding a glorious view of the valley below. An ascent to this point will well repay us, ere we start on our trip toward Upper Mauch Chunk.

But first, what say you to a run up to the Glen? It is only two miles away, and here comes a train that will take us up there in a few minutes. So jump aboard. We whiz away up the river bank, the thickly built street on our left, the river on our right; presently the houses end, and an embankment takes their place in our vision. Then we pass the coal shutes, rattle along under the base of Mount Pisgah, cross the river to East Mauch Chunk, shoot through a tunnel and over a bridge, and here we are at the pretty depot guarding the entrance to

---

Scale of distances of stations on the Central Rail Road of New Jersey, above mean low tide water.

Pier 14 North River	10.5	North Branch	87.1
Elizabeth	31.3	White House	178.1
Roselle	79.8	Lebanon	296.0
Cranford	74.6	Annandale	346.1
Westfield	130.3	High Bridge	331.2
Fanwood—new	158.1	Glen Gardner	450.7
Plainfield	107.9	Junction	508.5
Evona	78.5	Asbury	439.2
Dunellen	56.1	Bloomsbury	336.2
Bound Brook	31.4	Springtown	303.7
Finderne	81.5	Greenwich	262.3
Somerville	61.9	Phillipsburg	222.2
Raritan	72.8		

GLEN ONOKO.

It is a wild and tangled spot this to which our quest for the romantic has brought us. We are completely shut in by the mountains; for on either side the valley makes an abrupt turn, leaving us as it were enclosed in a deep basin. And before us is a gorge, or glen, from which a noisy waterfall comes babbling, foaming, plashing out to swell the Lehigh. That is Glen Onoko.

A few steps along a shaded pathway bring us to its portals, but not until we have had opportunity to slake our thirst at a wayside spring where the cool water spurts up in a stream of an inch in diameter, and to a height of at least two inches above the ground. To our left is Sentinel Rock, and a short distance beyond, Hidden Sweet Cascade.

Now we commence the ascent, shaded all the way. We pass, in turn, Entrance and Crystal Cascades, and, looking up, see above us a rustic bridge spanning the mountain stream. Upon this our path soon brings us, and we find ourselves face to face with Moss Cascade. At its base is a limpid pool known as the Lovers' Bath, while overlooking it are two immense boulders to which has been given the name of the Pulpit Rocks.* Now onward and upward we go again. The road is steep and trying, but there are new charms at every turn to repay us for our trouble. Here we cross another bridge and view the Spectre Cascade, deriving its wierd name from a resemblance, real or fancied, to the figure of a woman in white. Now we are in the Heart of the Glen, and, looking far up the vista, catch a sublime view of Chameleon Falls,

---

* The author acknowledges his indebtedness to the columns of that complete and attractive publication the *Monthly Souvenir*, (J. Lynn, Publisher), for much of his information relative to this locality.

and the Falls of Onoko, the former over forty, and the latter nearly ninety feet in height.

Ah! here is a stairway ingeniously hewed on the trunk of an immense hemlock. Then beyond we reach Sunrise Point, and, for the first time, gain a commanding view, as well as an idea of the altitude which we have reached. Looking down from this eyrie we recall the lines from the Lady of the Lake.

> "From the steep promontory gazed
> The Stranger, raptured and amazed.
> And 'What a scene were here,' he cried,
> For princely pomp, or churchman's pride!
> On this bold brow a lordly tower;
> In that soft vale, a ladie's bower;
> On yonder meadow far away,
> The turrets of a cloister gray.'"

After a feast of scenery which will have surprised and delighted even the most stolid, we resume our upward journey, pass Terrace Cascade, and reach Cave Fall, so called from its proximity to a rocky recess, which the Indians are said to have frequently sought as a place of concealment. Then our path leads us by a lumbermen's cabin to the Summit and to Packer's Point, where amid an extended view of the surrounding country, we have reached the climax of Glen Onoko's beauties.

Yet, so admirable are the arrangements for travel to and from the Glen, that in two hours after starting thither we are back again at the Mansion House, more than pleased with the charming scenes which we have visited.

But Mauch Chunk's grandest attraction yet awaits us Let us now start out on

A RIDE OVER THE SWITCH-BACK.

But what is the Switch-back, and whence its singular name? Briefly this. In 1796 one Philip Ginter, a pioneer in this region, accidently discovered the existence of immense beds of anthracite coal in the adjacent mountains. In 1818 the Lehigh Coal and Navigation Company was organized to develop these rescources. Their first problem was to solve the difficult question of transportation for coal from the mines to the river. There were the mines, inexhaustible in wealth, but buried in the heart of the mountain; there were the river and canal flowing at the mountain's base, nine miles distant, avenues of connection with all the populous world without. The problem was to connect the two, so science and enterprise joined hands to solve it. First the coal was carted by mule teams; but this tedious and expensive method was obviated in 1827 by the construction of what was called the Switch-back or Gravity railroad, running on a descending grade from Summit Hill to the river. Cars descended on this by their own gravity, carrying with them the mules which were to drag them back.

In 1844 the mule system was entirely abandoned by the erection of inclined planes on Mount Pisgah and Mount Jefferson, up which the cars are drawn by steam to the required elevation. And since that time the ride over these planes, and back over the Gravity road has been annually more popular among tourists, until now it has become an inseparable feature of a visit to Mauch Chunk.

Is it safe, do you ask? The best evidence of its safety is the fact that "in all the years that this enterprise has been in operation, not a single passenger has met with an accident going up this mountain." But come, let us see for ourselves.

Here is a carriage awaiting us, though we can walk to the plane in ten minutes, if we prefer. But there are some choice views by the carriage road, and we can take the foot path on our return. So we drive up Susquehanna Street, turn off to the left up Broadway, a closely built business street, crowded in curiously between the hills, and presently commence ascending the mountain slope. Ah, now the picture commences. Look down into the gorge behind us, and what a magical surprise awaits us. Here is the entire town below us, the river flowing before it, the engines whizzing by on either side, while standing out, prominent above all, is the symmetrical outline of St. Mark's Episcopal church, a feature indelibly impressed upon the memory of every one who has seen Mauch Chunk. And at every step that we advance the view seems to present some new and more beautiful aspect. Now we are at Upper Mauch Chunk. Ah! what is that? a railroad track? and see, there comes a car whizzing along with nobody in it. That is an excursion car returning by itself from Summit Hill to the foot of the plane. Just here we turn off at a zig-zag to the right, drive along the summit of the hill, pass the old cemetery, situated on a shaded bluff overlooking town, valley and river, and presently behold us at the foot of Mount Pisgah.

The view even here is fine enough to satisfy any reasonable sight-seer. But above us, rising at the rate of about one foot in three, is the plane, double tracked, and 2322 feet long, and when we shall have reached its summit we shall be 864 feet higher than we now are. So we will take our seats in the car, and start on our upward journey. The safety car takes its place behind us, the great heavy iron bands which extend from end to end of the plane, and form the medium

MT. PISGAH AND PLANE.

of our motive power, commence to move, and off we go, up, up, up. What a novel sensation it is; how steep the descent looks behind us; how far off the summit seems to be. This must be much the experience of a balloonist, when he first casts loose from his anchorage. Now we look over the tree tops, and every second the vista widens and widens below us. What were mountains a moment ago have dwindled now into mounds, or seem level with the landscape around them. Now, we pass the downward bound car which meets us midway. Now, we really begin to realize our immense elevation. The foot of the plane looks far, oh how far, below us. And, as we wonder and gaze in admiration, we are carried into the engine house at the summit, and the ascent of Mount Pisgah is achieved. We are now 1500 feet above tide water. Before us is a trestle bridge spanning a wild ravine, and passing over this we alight for a few moments and follow a winding pathway leading to a still higher point, the Pavilion. Here, from the summit of an observatory, which might well be named Tip Top, we gain a view as lovely as mortal eye ever gazed on, a view which justifies the assertion that we have found the "Switzerland of America." Away to the southeast is Lehigh Gap, and peeping through it, sixty-five miles distant, the rounded blue of Schooley's Mountain. Farther north is Wind Gap, and then following the horizon around we see a mingled panorama of blue hills and green forests bewildering in its extent and grandeur. No pen can do justice to this scene; no canvas can truly portray it. To see it is alone to appreciate it.

> "So wondrous wild the whole might seem,
> The scenery of a fairy dream."

Resuming our seats, our car by the force of gravity shoots us along the mountain side a distance of six miles, (a descent of 302 feet), to the base of Mount Jefferson, where a second plane 2070 feet long and 462 high, awaits our ascent. Again we enjoy the novelty of an aerial ride, and again we look down upon the landscape dwindling beneath us. Now we take another mile of gravity riding, (descending 45 feet), and—but what is this? a village? Yes, Summit Hill, a mining town of about 2000 people, built here on the mountain top, 975 feet above the Lehigh. A curious place it is, with rambling streets full of old buildings, among which we see a stone arsenal with loopholes for riflemen, and castellated towers. Here are stored the arms for a company of state militia, stationed here to suppress disorders. Yonder is the original switch-back railroad, from which the present one derived its name, and by which coal is brought up from the mines in Panther Creek Valley beyond. And here, too, is "the burning mine," which caught fire thirty-one years ago and has been ever since burning in its subterranean depths with a fiery heat, searing and blighting whole acres of ground on the surface above it.

But, the crowning delight of our ride awaits us. The return over the nine miles of descending grade to our starting point at Mount Pisgah's base. It only wants half an hour of supper time. Yet we can accomplish the distance in that time. Seated in the car, it is given a gentle push and off we go, down through long stretches of shaded roadway, down around wondrous curves, down along the edge of giddy precipices over which we look down upon tree tops far below; down under the shadows of great crags and walls of rock covered with luxuriant ivy, and still down, down, down, at a dizzy speed, and as if on the wings of the wind. Oh, this is exhilarating; the cool mountain air fans our brows, the sweet fragrance of the woods and wild flowers greets us; the entrancing scenery far below seems to shoot up to meet us as we momentarily near it. On we go with the speed of a race horse; nothing now dares obstruct our course. Ah, there is Mauch Chunk again, like a toy village in the distance below. With what wondrous rapidity we approach it. Faster

MANSION HOUSE

Mauch Chunk, Pa.

In a cool and pleasant location, within sight of all the depots.

FAVORITE RESORT OF TOURISTS.

PUREST WATER IN THE WORLD.

BALCONY SERENADES DAILY!

Elegant Rooms for Families.

E. T. BOOTH, Proprietor

---

and faster still we rush on. Yes! there is the carriage road, there the old cemetery, and, before we know it, our invisible steed has come to a halt at the base of the plane whence we started.

There, reader, ends the ride over the Switch-back Railroad. Once having enjoyed it, you will never forget it. It may be safely said that there is no jaunt in all America, or perhaps in the world, that equals it in grandeur and exhilarating effect.

So, retracing our steps by the foot path to the village, and the Mansion House, in the cool of this summer evening, we recall with joyous wonder the scarcely yet realized beauties of the ride, and thank the kindly fortune which prompted us to visit, sometime in our lives, Mauch Chunk and Glen Onoko and the Switch back Railroad.

Nor need our journey end here; for following our line still further northward, and skirting the Lehigh's waters toward their course, we may, after bidding adieu to Mauch Chunk and its pleasant memories, reach in turn White Haven and the bold declivities beyond it, Wilkesbarre and the lovely Wyoming Valley, prosperous Scranton in the heart of the Pittston coal region, and thence continue on to Niagara Falls, Saratoga, Watkins Glen, Sharon Springs, Albany, Lakes George and Champlain, and the Canadas. What more fascinating or delightful summer excursion could even the most *blase* tourist desire?

NEWARK AND NEW YORK RAILROAD.
(See page 7.)

After leaving Communipaw, we whiz across a strip of meadow, and stop for a moment at the signal station at

LAFAYETTE,
(15 min. 11 trains each way daily.)

the name originally pertaining to this whole section, which is now a portion of Jersey City. Just beyond the depot we find ourselves running along a high embankment, and presently cross in turn the Morris Canal and the Bergen Plank Road, each of them spanned by substantial bridges of iron. But suddenly the land rises on either side of us, and almost ere we know it we are in the Bergen Hill Cut, and look down from the car windows on our left upon a cut fully as deep again, an immense excavation in solid rock, destined on its completion to afford our line a much easier grade of roadway to the meadows beyond the hill. Now, as the whistle blows, we come to a stop at

BERGEN AVENUE,
(17 min. 27 trains each way daily.)

a magnificent highway, extending to the heart of the city, and forming the main avenue of communication in this section of it. The short time required to reach this station from the city, its frequent trains, and the number of unusually attractive building sites within its immediate vicinity, render it noticeable as the most important point we have yet reached on the line. (See advertisement of Woodward & Sherwood, page 7). At

WEST BERGEN
(19 min. 17 trains each way daily.)

we find ourselves on the shore of the Hackensack River, here grown to a considerable width and depth. Now we cross the Morris Canal again, shoot out upon the long trestle bridge, and, in a moment or two have passed over it, over the neck of green meadow land, dividing the Hackensack and Passaic, over the Passaic itself, and are whizzing on to BRILL'S JUNCTION, where the Newark and Elizabeth Branch comes in. Now we are already in Newark, and our first station within its limits is at

EAST FERRY STREET,
(29 min. 21 trains each way daily.)

where already numerous indications of city conveniences appear, in the way of gas lamps, curbed streets, sidewalks, hydrants, and substantial brick rows. Here our track is above the level of the city, necessitating a bridge at every few hundred feet. It is a peculiar advantage that this line has in passing through the entire city *above the level of its streets*, thereby obviating the delays and collisions incident to other lines *on* the level. In so constructing it the Company have simply anticipated what must in a few years become a general system of railroad transit through cities. We come next to

FERRY STREET,
(31 min. 27 trains each way daily.)

where we are in a thickly settled portion of Newark, and where our attention is especially attracted by the elegant and spacious brick depot, erected for the accommodation of residents in this section of the city. The gas works may also be seen to the right of the railroad at this point.

We now find ourselves carried along over the level of the housetops, and can, if curious, look down into many a second story window, or back dooryard. Now we find ourselves in a moment or two at the

BROAD STREET DEPOT.
(35 min. 27 trains each way daily.)

Conveniently located in the heart of the business portion of Newark, and with horse-cars in waiting to convey us to any part of the city.

SOUTH BRANCH RAILROAD.
(See page 52.)

Upon leaving Somerville, we first cross the Raritan, and passing through a fine, open farming country, come to

RICEFIELD.
(1 hour, 55 min. 4 trains each way daily.)

A small village, once known as Royce Field, lying southeast of the road. One gains here a fine view of the Pickles Mountain, on the northwest. Two and a half miles beyond is

FLAGTOWN.

(2 hours and 2 minutes. 4 trains each way daily.)

A depot where passengers alight for the village of the same name, about half a mile to the south.

NESHANIC.

(2 hours, 10 min. 4 trains each way daily.)

Situated on the south branch of the Raritan, is, next to Flemington, the most important point on this branch. Ninety-three thousand baskets of peaches were shipped from here during the last season. About the depot may be seen many tasteful dwellings, while the village proper, at the cross-roads, half a mile south, is quite a center of population.

Following the line of the south branch, we come next to

THREE BRIDGES,

(2 hours, 20 min. 4 trains each way daily.)

a village boasting a Church and a Post Office, but chiefly important as an outlet for the farming country about it.

Beyond the village we again cross the south branch, enter Hunterdon County, and suddenly catch a glorious view of a wide-spread, fertile valley, in the midst of which is seen the respectable county town of

FLEMINGTON.

(2 hours, 30 min. 4 trains each way daily.)

Of this place we may to-day with correctness repeat the description by a writer of upwards of thirty years ago, who refers to it as "principally located on a single street, on which are many handsome dwellings; and the general appearance of the place is thriving and cheerful." It being the Hunterdon County Seat, we find here a stone Court

House, and County Offices. There are three or four churches, good schools, some fine stores, and an estimated population of from three to four thousand. Gas lamps and hydrants give a sort of metropolitan air to the streets, and two steamers are provided for the extinction of fires.

NEWARK AND ELIZABETH BRANCH.

This branch, which passes across the Newark meadows, and is regularly run as a local line between the cities of Newark and Elizabeth, furnishes to the people of the former city a close connection from and to nearly all the trains on the main line, and, judging from the amount of travel passing over it, is a great public convenience.

BRANCH TO
P'TH AMBOY, S. AMBOY & LONG BRANCH.

The manifest need of an "all rail" connection between New York and its most popular sea-side watering-place determined the Company in the construction of a branch line to the latter, and the work now nears completion. The line diverges at Elizabethport; runs thence to Woodbridge and Perth Amboy; crosses the Raritan to South Amboy by a substantial and handsome bridge of 3,000 feet, and thence runs through Red Bank to Long Branch. It is expected to open the road to Long Branch next season, and to Perth Amboy this summer. Already the results of direct communication by rail with New York, and by bridge with its sister town South Amboy, are visible at Perth Amboy in the enhanced value of property, and the constant accessions to its population.

A CLUSTER OF GOLDEN OPINIONS
FOR THE
BRADBURY PIANO

ITS ADAPTATION TO THE HUMAN VOICE as an accompaniment, owing to its peculiar sympathetic, mellow, yet rich singing qualities and powerful tone.

☞ From present acquaintance with this firm we can indorse them as worthy of the fullest confidence of the Christian public. We are using the Bradbury Pianos in our families, and they give entire satisfaction.

Persons at a distance need feel no hesitation in sending for their illustrated price-list, and ordering from it, or to order second-hand Pianos. They are reliable.

Mrs. U. S. Grant, Wash., D. C.	Attorney-Gen. Williams
Chief-Justice Chase, Wash., D. C.	Rev J M. Walden, Chicago.
Vice-Admiral D. D. Porter, Wash., D. C.	Rev R M Hatfield. Cin
	Rev L B Bugbie, Cin.
Hon. Columbus Delano, Wash., D.C.	Dr. J. M. Reid, N Y.
P. M. General Creswell, Wash., D.C.	Dr. C. N. Sims, Balt., Md
Robert Bonner, N. Y.	Dr. H. B. Ridgaway, N. Y.
Grand Central Hotel, N. Y.	Philip Phillips, N. Y.
St. Nicholas Hotel, N. Y.	Rev. Alfred Cookman, N. Y.
Metropolitan Hotel, N. Y.	Rev. John Cookman, N. Y.
Hon. J. Simpson, M. P.	W. G. Fischer, Phila , Pa.
Bishop M. Simpson, Phila	Chaplain M'Cabe.
Bishop E. S. Janes, N. Y.	Rev. A. J. Kynett, D. D.
Wm Morely Punshon	Rev. Daniel Curry, D. D.
T S Arthur, Phila	Theodore Tilton, N. Y.
Dr. John Chambers, Phila.	Dr Daniel Wise, N. Y.
Rev S. W. Thomas, N. Y.	Rev. W. H. Ferris, N. Y.
Rev. I. W. Wiley, Chicago.	Rev. Dr. Fields, N. Y.
Rev. J. S. Inskip, N. Y.	Sands St. Church, Brooklyn.

The best manufactured ; warranted for six years. Pianos to let, and rent applied if purchased : monthly installments received for the same. Old pianos taken in exchange; cash paid for the same. Second-hand pianos at great bargains from $50 to $200. Pianos tuned and repaired.

ORGANS AND MELODEONS

To Sabbath-Schools and Churches supplied at liberal discount. Send for illustrated price-list.

F. G. SMITH & CO.,
Late Supt. for and successor to WM. B. BRADBURY,
127 Broome Street, New York.

FREEBORN G. SMITH.   H. T. M'COUN.

THIS COMPANY IS PREPARED TO FURNISH

MAPS OF PROPERTY,
(See page 45)

CUTS OF BUILDINGS, Etc.
(See page 60)

BILL OR LETTER HEADS,
(See above)

Relief Plates for Newspaper, Book and Catalogue Illustrations,

AND

ENGRAVING WORK IN GENERAL,

at prices which average about

ONE-HALF OF RATES CHARGED FOR WOOD CUTS.

Their work is engraved in very hard Type Metal, by a new chemical process, direct from all kinds of Prints, Pen-and-Ink Drawings, Original Designs, Photographs, &c. This process is in many respects vastly superior to wood engraving. The plates have a printing surface as smooth as glass, and the lines are deeper than those of hand-cut engravings. We guarantee all our plates to print *absolutely clean and sharp* on either wet or dry paper, and on any kind of press where type or wood cuts can be printed. The attention of manufacturers proposing to issue Illustrated Catalogues is particularly invited.

L. SMITH HOBART, Pres't. **J. C. MOSS, Sup't.**

D. I. CARSON, Gen'l Agent.

New Family SINGER Sewing Machines

DURING 1872

The Singer Manufacturing Company sold 219,758 Machines.

Wheeler & Wilson Manufacturing Company	" 174,088	"
Howe Machine Company (estimated)	" 145,000	"
Grover & Baker Sewing Machine Company	" 52,010	"
Domestic Sewing Machine Company	" 49,554	"
Weed Sewing Machine Company	" 42,444	"
Wilcox & Gibbs Sewing Machine Company	" 33,639	"
Wilson Sewing Machine Company	" 22,666	"
Amer. B. H. O. & Sewing Machine Company	" 18,930	"
Gold Medal Sewing Machine Company	" 18,897	"
Florence Sewing Machine Company	" 15,793	"
B. P. Howe Sewing Machine Company	" 14,907	"
Victor Sewing Machine Company	" 11,901	"
Davis Sewing Machine Company	" 11,376	"
Blees Sewing Machine Company	" 6,053	"
Remington Empire Sewing Machine Company	" 4,982	"
Keystone Sewing M. Co.	" 4,262	"
Bartlett Reversible Sewing Machine Company	"	"
Bartram & Fanton Manufacturing Company	" 1,000	"
Secor Sewing Machine Company	" 1,000	"
	" 311	"

O. T. HOPPER.   **O. T. HOPPER & CO.,**   H. V. D. SCHENK

New Jersey Agency for The Singer Sewing Machines,

No. 766 Broad Street, Newark, N. J.

O. T. HOPPER & CO., 163 Broad Street, Elizabeth.

POPE BROS., Front Street—Plainfield.

WM. C. SHAFER, Main St.—Somerville

A. CARTER, Phillipsburg, opposite D. & B. R. R. Station.

J. S. & J. N. CLARK, Lebanon.

WM. H. FULPER, Main St.—Flemington

THE PATENT ARION
PIANO-FORTE

EXCELS ALL OTHERS IN
TONE AND DURABILITY!

The Arion Piano-Forte contains in its construction four valuable patented improvements that make it MORE DURABLE than any other piano. The sales of these pianos have increased over SIX HUNDRED per cent. in the past three years. The Arion Pianos are used exclusively by the N. Y. Conservatory of Music because of their unequalled Tone and great Durability. Great inducements to cash purchasers. WRITE for ILLUSTRATED CIRCULAR, and MENTION WHERE you saw this NOTICE. Address,

ARION PIANO-FORTE CO.
No. 5 East 14th Street,
NEW YORK.

No. 1308 Chestnut St., Philadelphia.

No. 211 State St., Chicago.

No. 214 North 5th St., St. Louis.

No. 20 O'Farrell St., San Francisco.

☞ Pianos sold on easy Monthly Payments and old pianos taken as part pay.

CENTRAL NEW JERSEY
Land Improvement Co.

HAVE FOR SALE

Dwellings, Lots and Villa Sites,

AND

LAND BY THE ACRE

SUITABLE FOR

☞ **Manufacturing Purposes!**

—AT—

NEWARK,	BAYONNE,	BERGEN POINT,
ELIZABETH,	CRANFORD,	WESTFIELD,
FANWOOD PARK,	PLAINFIELD.	EVONA,
DUNELLEN,	SOMERVILLE,	RARITAN AND
CLINTON,	BLOOMSBURY,	PHILLIPSBURG,

Apply to

Or

A. D. HOPE.

ELSTON MARSH,

119 Liberty Street, New York.

www.ingramcontent.com/pod-product-compliance
Lightning Source LLC
Chambersburg PA
CBHW031604110426
42742CB00037B/1097